COLD WAR HOTHOUSES

COLD WAR HOTHOUSES

INVENTING POSTWAR CULTURE FROM COCKPIT TO PLAYBOY

BEATRIZ COLOMINA,
ANNMARIE BRENNAN,
AND JEANNIE KIM,
EDITORS

PRINCETON ARCHITECTURAL PRESS

Published by
Princeton Architectural Press
37 East Seventh Street
New York, New York 10003

For a free catalog of books, call 1.800.722.6657.
Visit our web site at www.papress.com.

Printed and bound in Canada by Friesens
07 06 05 04 5 4 3 2 1 First edition

Publication of this book has been supported by a grant from the Graham Foundation for Advanced Studies in the Fine Arts.

Editing: Nancy Eklund Later
Editorial Assistance: Linda Lee
Design: Deb Wood

Special thanks to: Nettie Aljian, Janet Behning, Megan Carey, Penny (Yuen Pik) Chu, Russell Fernandez, Jan Haux, Clare Jacobson, John King, Mark Lamster, Katharine Myers, Jane Sheinman, Scott Tennent, Jennifer Thompson, and Joseph Weston of Princeton Architectural Press—Kevin C. Lippert, publisher

Frontispiece: "The Bride's House of 1956," House Beautiful 86, no. 5 (May 1956), back cover.

Library of Congress Cataloging-in-Publication Data

Cold war hothouses : inventing postwar culture, from cockpit to playboy / Beatriz Colomina, AnnMarie Brennan, and Jeannie Kim, editors.
p. cm.
Includes bibliographical references.
ISBN 1-56898-302-6 (pbk. : alk. paper)
1. Material culture—United States—History—20th century. 2. Technological innovations—United States—History—20th century. 3. United States—History—1945- 4. United States—Social conditions—1945- 5. United States—Social life and customs—1945-1970. I. Colomina, Beatriz. II. Brennan, Annmarie. III. Kim, Jeannie.

GN560.U6C65 2004
306'.0973'0904—dc22 2003027468

for Ignasi de Solà-Morales, in memoriam

CONTENTS

ACKNOWLEDGMENTS

The editors and authors would like to thank the Graham Foundation for Advanced Studies in the Fine Arts for its support.

We are grateful to Stan Allen, Dean of the School of Architecture, and Professor Mario Gandelsonas at Princeton University for their support of this endeavor. Assistance with acquiring and organizing photographs was graciously provided by Audio Visual Curator Amy Herman. We would also like to thank Francine Corcione, Linda Greiner, Cynthia Nelson, Rena Rigos, and Erik Johnston. In addition, we were often aided by the helpful and knowledgeable staff of the Princeton School of Architecture Library, directed by Frances M. Chen.

BEATRIZ COLOMINA, ANNMARIE BRENNAN, AND JEANNIE KIM

COLD WAR HOTHOUSES

BEATRIZ COLOMINA

Tailfins, Saran Wrap, ball point pens, Slinky toys, Tupperware, highways, aluminum, Hula Hoops, food blenders, plastics, Barbie dolls, dishwashers, credit cards, office systems, picture windows, bikinis, fast food, TV dinners, drive-in cinemas, play rooms, window air conditioners, satellites, missiles, bomb-shelters, tranquilizers...: the artifacts of Cold War America. By-products of the war effort, they represent the period as if in a constellation, the result of an atomic explosion, each one of them a fragment for detached analysis or speculation, able to bring light into the period; as if an archeological dig where a piece of a jug may help to understand an entire culture, its habits, degree of development, artistic tendencies and so on. By looking closely at some of these remnants we see the cold war period in all its complexity. An archeology of our own period, a time that still haunts us.

This collection of fragments is not complete, and how could it ever be? It is simply an assemblage of pieces picked up for analysis by a group of doctoral candidates at a series of seminars and workshops at the School of Architecture at Princeton University, conducted between 2000 and 2002. The theme of the seminars was postwar America, a subject neglected in architectural research until recent years, when several conferences, exhibitions, and articles started to open up the field. At the time, my own research on the period had focused on the impact of World War II on architectural discourse, the new roles played by architects who had been intimately involved in the war effort and the recycling of military technologies, materials, and attitudes to the reorganization of postwar

Tupperware products displayed on a highway

space and lifestyle: in other words, the redefinition of the architect and architectural design by the war.[1] This research was used to launch the seminar "Cold War/Hot Houses" in the fall of 2000, but participants were then encouraged to come up with their own topics of study. The new topics tended to be less about the work of architects and more about the wide-ranging transformation of space in the postwar years. Instead of talking about designers, buildings, architectural details, designer furniture, master plans, professional publications, and the like, the research paid close attention to popular magazines and books, advertisements, movies and television programs, governmental initiatives, and developers schemes. The seminar tested the idea that the postwar period no longer celebrated the heroic figure of the architect transforming the spatial order, even though most architects were still modeling themselves as heroic. The real changes were going on elsewhere. Objects of everyday life involved more radical transformations of space than the most extreme architectural proposals. Indeed, the most radical architects were those who were able to understand and respond to these cultural and technological shifts.

In that sense the Cold War itself was a hothouse, breeding new species of space, a new organizational matrix. Hence the title *Cold War Hothouses*, meaning not simply the effect of the Cold War on house design but all of the new forms of domesticity that emerged during the period and that in many ways we still occupy today.[2] The initial seminar focused on domesticity: children, beds, bomb shelters, lawns, TV sets, picture windows, glass houses, appliances, plastics, toys, food, etc. The following year we expanded to the urban and territorial scale: drive-in theaters, shopping malls, highways, standardized IQ testing, the national park system, the office cubicle, etc. Soon we realized that scales had become conflated, that everything in the postwar age was domestic.

The drive-in cinema, for example, turned out to be as domestic a space as the suburban home, with the car acting as a new kind of living room on wheels, where teenagers made out, the housewife carried out domestic chores such as sewing or snapping peas, children played, babies slept, the convalescents rested, the family ate... all within the confines of the automobile, while looking at a screen that, to make the point literal, sometimes had the shape of a gigantic TV set. As David Snyder demonstrated in his seminar paper, the drive-in was promoted as a living-room, bedroom, and kitchen, where you could take the kids, dress as you wish, have a conversation, smoke, make noise, seduce, and perhaps watch a film.[3] Public space could only be sold by offering it as a form of domestic privacy not so different from that experienced in the suburbs, where each living room was exposed through a picture window, and houses sat side by side with almost identical ones, all tuned to the same TV program; as Jack Kerouac put it in his 1958 *The Dharma Bums*,

> Rows of well-to-do houses with lawns and television sets in each living room with everybody looking at the same thing and thinking the same thing at the same time.... House after house on both sides of the street with the lamplight of the living room, shining golden, and inside the little blue square of the television.[4]

Even the monumental national park system was understood as an extension of domestic space. As Jeannie Kim argues, it was

Snapping peas at the drive-in, circa 1950

designed to be experienced by the family within the interior of the car or on trails specifically designed to accommodate women in high heels. The visitor centers, with their open plans and flat roofs, had the architecture of a modern suburban house, and the drive-in camping sites resembled the lay-out of a suburban neighborhood, complete with cul-de-sacs. The great outdoors was transformed into a comforting domestic interior, a reassuring space, with dependable trails, labels, narratives, photo opportunities, and orientation movies (in fact, the whole thing was experienced like a film, a highly controlled outdoor museum). If the nineteenth-century park offered a respite from the ills of urban life, the national park system, experienced on wheels in the containment of the family unit, reproduced the suburban life common to most visitors. The experience was offered as a kind of group therapy for a still-traumatized nation.

The vast space of the highway also turns out to be a domestic enclave. The Beat Generation, "on the road," tracing and retracing the U.S. highway system, moved the fulcrum of literature, as Roy Kozlovsky argues, from time and history to space. Kerouac made a calculated decision to leave home and hit the road in order to write. For him, writing equaled driving and the car, the vehicle of the twentieth-century novel. And yet the movement across the U.S. recorded in his epic 1957 novel *On the Road* is always punctuated by houses: the interiors they depart from, the stops along the way, are organized by multiple interiors and domestic scenes, no matter how unconventional. There is hardly ever a description of public spaces, with the exception of some bars, and little TV or film. The car itself is once again an interior on wheels, a space to eat and sleep and have sex, the TV set replaced by the ever-changing view through its windshield. The trips, crossing and recrossing America, it could even be argued, have the structure of TV reruns of prime-time soap operas and situation comedies, with the character of Dean Moriarty endlessly falling in and out of relationships with multiple girlfriends, and multiple wives, with friends, with his lost father, and with his companion Sal Paradise, who keeps waiting for his aunt to wire him money. Dean and Sal don't move to find new people but to find the same ones again and again. If this mobility and homelessness is, as Kozlovsky argues, the alternative to suburban boredom in *On the Road*, written on a long scroll of architectural drafting paper, Kerouac, who had worked for the construction of the Pentagon and whose trip is being paid for by the GI Bill, creates a space as domestic as the suburban house he is trying to escape. Indeed, the whole highway system becomes one small domestic world.

If public space was privatized, domestic space was publicized, not just on view (TV was already advertised during World War II as the "biggest window in the world") but on the move, mobilized: the TV set was placed on wheels, the walls became partitions, and the housewife seemed to be always in a hurry with a barrage of conveniences, push button devices, and appliances, designed to save her time: quick mixes, fast food, Tupperware, blenders, dishwashers, washing machines.... This new kind of mobility and efficiency had to do with the war. Not only was her "push button" equipment coming from the same factories that made guided missiles,[5] but the house itself was defending the nation. The housewife had become a soldier on the home front; the kitchen, the command post from where she not only controlled the domain of her living space but was

"The Bride's House of 1956," *House Beautiful,* May 1956, back cover

Tupperware home party, Sarasota, Fla., 1958 (left); Richard Nixon and Nikita Khrushchev, with plastic bowls handed out as souvenirs at the American National Exhibition in Moscow, 1959 (right)

purported to defend the nation. The suburban house, equipped with every imaginable appliance, projected the image of the "Good Life," of the lifestyle of prosperity and excess that was the main weapon in the Cold War. This became evident in the 1959 "Kitchen Debates" between American Vice President Richard Nixon and USSR premier Nikita Khrushchev, when appliances rather than missiles were identified as the sign of strength of a nation. Politics had moved to the domestic space—or, more specifically, to the kitchen of a suburban house put up by a Long Island builder and furnished by Macy's for the American National Exhibition in Moscow's Sokolniki Park. Their debate was broadcast to the world. For Nixon, American superiority rested on the ideal of the suburban home, complete with modern appliances and distinct gender roles. He proclaimed that this suburban home, with its "many different kinds of washing machines" to choose from, represented nothing less than American freedom.[6]

The organization of the domestic space during the Cold War years and even the language used to describe it echo this movement of military logic into the private sphere. The all-plastic Monsanto House of the Future, displayed at Disneyland between 1957 and 1968, and here analyzed by Stephen Phillips, had a "control tower" in the kitchen where the housewife could not only operate everything in the house by push button but where she could, in the words of a reporter, "maintain surveillance over three of the four rooms in the house." Sent to the World's Fair in Brussels in 1958 as part of the "Face of America" Pavilion, the Monsanto project constitutes one of the multiple examples of an American house deployed as weapon in the Cold War. The face of America during the Cold War was a hot house.

Communism fought with washing machines and food mixers. The conflation of public and private had become evident in the official policies of the period. McCarthy, the Federal Bureau of Investigation, and the fear of Communism turned everybody into a spy of everybody else, particularly neighbors and coworkers. Official investigations obsessed mainly on who was meeting in whose house to discuss what with whom. In the 1954 McCarthy hearings, secrets were exposed (or kept) live for thirty-six days in front of a TV audience of 20 million.

Two years later, *Playboy* exposed a different kind of private secret for a mass audience. While the magazine thought of itself as poised against conventional domesticity—a "pornotopia," in Beatriz Preciado's analysis, an escape from the stifling suburbia of 1956—in fact, it represented another form of domesticity. In some ways, with its designer furniture, its mobile elements, its radical transparency, it was the exacerbation of the postwar house fantasy, a very hot house indeed. *Playboy* endlessly presented hyper-designed domestic interiors and ideal scenarios for bachelors, making it acceptable for men to be interested in architecture and interior decoration. Hugh Hefner's *Playboy* mansion became identified as the paradigmatic playboy house. As Hefner himself put it when interviewed years later, he wanted the mansion to be a "dream house."[7] Its mid-century modern interior architecture and furniture included glass walls, Saarinen chairs, light partitions, and every single appliance imaginable. The entire space was controlled by push button, from a central control panel—not so different from the push button fantasies of futuristic houses at world's fairs.

If prostitution is urban and depends on the streets of the city, *Playboy* is domestic and designed for the age of the green-lawn suburbia. *Playboy* played in a domestic space for a domestic audience.

The houses Hefner created and circulated through his magazine were even broadcast on TV. Erotica in the age of suburbia is the fantasy of the girl next door, delivered to one's bedroom through the media.

Symptomatically, the dually domestic and public realm of the *Playboy* mansion is also Hefner's office. Most of the furniture and partitions came from the office world, acting as flexible props that can be endlessly reorganized to sustain multiple scenarios. Domestic space absorbs the new logic of office space, which in turn came from the military. As Branden Hookway demonstrates, the military strategy of endlessly rearrangable scenarios infiltrated the postwar office environment; or rather, the office space with its systems furniture was of military design. The logic of the cockpit simply became the logic of the cubicle. The mentality of tactical operations, of rapidly reorganizing resources to confront ever-changing scenarios, became the most effective business strategy. The office worker fused dynamically with the office space in a new mode of active intelligence that pervaded the postwar landscape.

Take the FORECAST program of the American Aluminum Company of America (Alcoa), for example, which enlisted some of the most talented designers and architects of the period in its attempt to convert aluminum to peacetime use after World War II. The commission was not simply to create products using the material, but as AnnMarie Brennan has analyzed, to "forecast"—in the military sense of anticipating a situation—possible scenarios for its use. With the help of famous fashion photographers, the campaign consisted of a series of advertisements featuring new products and envisioning scenarios for their use. Not surprisingly, the products tended to be not just domestic (tables, picnic shelters, vacation houses), but folding, collapsible, portable, nomadic spaces. Lightweight and compact, as the material was touted, the aluminum house was the postwar soldier's house—demountable, temporary, on the go.

Even toys were military. Charles and Ray Eames's 1951 invention called *The Toy* had, as Tamar Zinguer argues, similarities with kites used during World War II, both as part of rescue kits and as targets used for gunners' practice. The 1949 Eames House itself was understood as a kite—a gigantic, lightweight, colorful, rearrangable toy, constructed of machined, off-the-shelf parts, remnants of the war industry. *The Toy*, a set of colored geometric panels designed for the amusement of adults and children alike, created "A Light, Bright, Expandable World Large Enough to Play In and Around," according to the label. A house, a world; *The Toy* was all scales, from domestic to planetary, collapsed into one open system.

The Eames's idea of design turns on the continuous arrangement and rearrangement of a limited kit of parts. Almost everything they produced can be rearranged; no layout is ever fixed. Not only *The Toy* but also the famous plywood cabinets, the House of Cards, the Revell Toy House, and the Kwikset House are all infinitely rearrangable. The Eames House provides a good example: not only was it produced out of the same structural components as the utterly different Entenza House (designed around the same time by Charles Eames with Eero Saarinen), but it was itself a rearrangement of an earlier version. After the steel had been delivered to the site, the Eameses decided to redesign the house by putting the same set of steel parts together in a completely different way than they had planned.[8] As Peter Smithson put it, "In the Eames house in Santa Monica, at the point the decision was made to use the same steel in a different way, that part of Eames' mind which worked best, direct-working assembly, took-over."[9]

Kits of parts were an integral part of the postwar culture. Could it be that the arranging and rearranging of fragments to make new forms and spatial structures had also to do with the war? Not just in the sense that the structures built that way are like military equipment—lightweight, demountable, engineered for rapid deployment, collapsible and able to adapt to any set of circumstances and uses—but because the very idea of putting together again and again a world out of a set of small fragments may have given people a sense of control over their environments in a world threatening to explode any minute. Play as therapy; architecture as "re-orientator" and "'shock absorber'": this is what the Eameses argued was the role of

"Family Utopia," from *Life* magazine, November 25, 1946

the house, a definition that is hard to conceive without thinking about the war: "The house must make no insistent demands for itself, but rather aid as a background for life in work . . . and as re-orientator and 'shock absorber.'"[10] If, according to the Eameses, "a good toy contained in it clues to the era in which it was produced,"[11] postwar playthings are symptomatic of the atomic era.

The entire Cold War culture blurs the distinction between work and play, business and entertainment, appliances and toys, buildings and dollhouses. Real estate companies used to give customers toy house sets to play with and envision the multiple possible arrangements of their future house. Manufacturing developed reconfigurable systems rather than fixed suites of furniture. The consumer was treated as an intelligent and playfully creative decision maker. In fact, every postwar artifact is a kind of toy, from Tupperware to the Space Program. American culture and its playthings: each one of them, seemingly unimportant, even frivolous, but each offering clues to the wider, darker era that produced them. Gathered together here like archeological finds, these minor objects of daily life construct a new kind of portrait of the Cold War. Each toy sets up a different kind of defensive play—play not as the escape from Cold War tension but as its very *modus operandi.*

The seminar out of which these studies grew was itself a hothouse for ideas, a kind of playground for serious play. The writers combined archival documents in new ways, playing with our assumptions to reconsider architect's role in a crucial historical period. This collection represents the thinking of an emerging generation of young scholars whose interests are already different from those of the generation before them and will no doubt change in the future. The work is exploratory, asking questions rather than answering them. By focusing on the spatial effects of some of the seemingly mundane products of the Cold War, it aims to open up both new ways of thinking about architecture and new ways of thinking about a period characterized by major infrastructural transformation in American life, a period that we may still be within, a hothouse we still occupy.

COCKPIT

BRANDEN HOOKWAY

The United States' World War II war effort can be broadly characterized as a professionalization of the war machine. This professionalization was driven by the increasingly important role played by logistics in modern warfare, from the gathering and processing of intelligence on the battlefield to the development and management of large-scale scientific projects. The military found that the key to its successful operation was continuous innovation in its technologies and organizational structures—its ability to transform itself, both in response to potential shifts in the geo-political context and in anticipation of the tactics these often-hypothetical situations would require.[1] To maintain a state of permanent institutional change, the military relied increasingly upon industry and academia, and one of World War II's great legacies was the growing interrelation of these three realms. For sociologist C. Wright Mills, this would be reflected in the unprecedented postwar transfer of personnel between the military and the private sector: whereas in 1900 "only three of the top three dozen army men ever went into business—and two of these were non-regulars," in the aftermath of World War II a mass movement of top generals into the corporate realm took place.[2]

At the same time, scientists who had participated in the war effort realized that much of the work done in wartime could easily be extended to industry and to society at large. For Vannevar Bush, who had left his position as a professor of electrical engineering at the Massachusetts Institute of Technology to oversee the U.S. World

Man-machine interaction in World War II aircraft design (opposite, top); Herman Miller, Action Office, circa 1968, fisheye view (opposite, bottom)

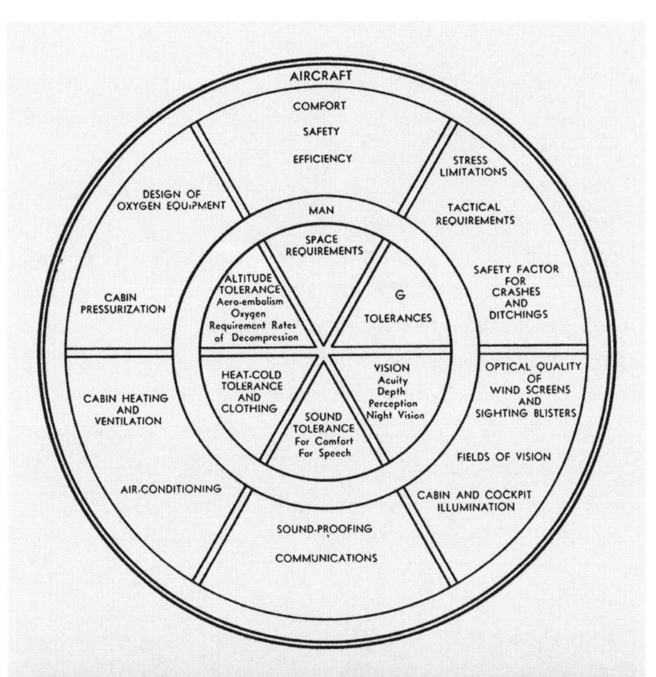
AIRCRAFT
COMFORT
SAFETY
EFFICIENCY
STRESS LIMITATIONS
TACTICAL REQUIREMENTS
SAFETY FACTOR FOR CRASHES AND DITCHINGS
OPTICAL QUALITY OF WIND SCREENS AND SIGHTING BLISTERS
FIELDS OF VISION
CABIN AND COCKPIT ILLUMINATION
SOUND-PROOFING
COMMUNICATIONS
AIR-CONDITIONING
CABIN HEATING AND VENTILATION
CABIN PRESSURIZATION
DESIGN OF OXYGEN EQUIPMENT
MAN
SPACE REQUIREMENTS
G TOLERANCES
VISION Acuity Depth Perception Night Vision
SOUND TOLERANCE For Comfort For Speech
HEAT-COLD TOLERANCE AND CLOTHING
ALTITUDE TOLERANCE Aero-embolism Oxygen Requirement Rates of Decompression

War II scientific effort, the problems encountered in a nascent post-war information economy would directly follow upon the data-management problems of mobilizing the U.S. war machine. In a 1945 article that appeared in both the *Atlantic Monthly* and *Life* magazine, Bush mused, "Man has built a civilization so complex that he needs to mechanize his records more fully if he is to push his experiment to its logical conclusion and not merely become bogged down partway there by overtaxing his limited memory."[3] If, for Bush, the complexity of war folded seamlessly into the complexity of civilization, then the protocols—organizational as well as technological—developed to mitigate this complexity would quickly become naturalized within civilian life. Like Memex, a computer-based content-association machine proposed by Bush featuring perhaps the first theorization of hypertext, these techniques would involve extending human ability, not only in terms of reach and physical power but also in terms of sensation, vision, memory, and, above all, *intelligence,* to which all other human qualities are ultimately subject. In this way, the main legacy of the professionalization of warfare with regard to the qualities and abilities of humankind can be said to be the systematization of intelligence.

The modern military claim upon intelligence extended well beyond the augmentation of individual ability to address what could be called organizational intelligence, or the way that organizations identify and correct their own inefficiencies and pursue their own aims. The unprecedented logistical requirements of modern global warfare, as recognized from the outset of World War I, demanded innovation at both the individual and organizational level.

Whereas prior to World War I and again between the two world wars, the U.S. military had relied upon a largely pre-modern organizational system of promotion through apprenticeship, the effort of staging global war forced the rapid rationalization of personnel selection and advancement. Faced with the pressing need to create a sophisticated army out of a civilian economy, military leaders turned to psychologists. Their work brought industrial and organizational theory to the problem of raising, training, and deploying armies as well as to the design of man-machine interfaces, most significantly in combat aviation. As with other areas of endeavor, military psychology found fertile ground in postwar culture, in consulting practices that sought to help industry adjust to the accelerated creation and dispersal of new technologies and to address through experimentation and design the critical socio-technical problem of the interface.

For American psychologist Kurt Lewin, whose work on group decision-making and social change in the 1930s and 1940s influenced the development of industrial and organizational psychology in both the U.S. and Britain, the advances that took place in the social sciences during World War II could be readily compared in significance to the creation of the atom bomb. The war provided the social sciences with both vital resources and a range of practical problems. Finding solutions to these problems would demand a unified, pragmatic, and ultimately professionalized approach. Instead of seeking only to understand human behavior, the discipline of psychology would need to become an applied science. For Lewin, the war amplified a shift already underway in the social sciences:

> Applying cultural anthropology to modern rather than "primitive" cultures, experimentation with groups inside and outside the laboratory, the measurement of socio-psychological aspects of large social bodies, the combination of economic, cultural, and psychological fact-finding, all of these developments started before the war. But, by providing unprecedented facilities and by demanding realistic and workable solutions to scientific problems, the war has accelerated greatly the change of social sciences to a new developmental level.[4]

In effect, the war drove the insertion of military psychology into all aspects of human life.

While the problems dealt with in military psychology varied greatly in type, ranging from the design of flight instrumentation to the development of aptitude tests to be applied to millions of men, a common approach to problem-solving emerged precisely at the intersection of experimental process and mathematical technique. Thus Lewin could write about the war as "integrating the social sciences" and about the application of techniques developed in the social sciences to "dynamic problems of changing group life."[5] What was inaugurated in the social sciences during World War II was a new form of instrumental rationality, one that no longer merely sought to reduce complex processes into their constituent parts. Instead, rationality took as its primary field of operations the *system*, or those processes that cannot be reduced from the material interactions that bring them into effect. As a result of this, the category of "individual"—a formal distinction that inherently casts the human being as raw material subject to training and discipline, to be molded into part of a larger organizational body—would thereafter cease to hold its singular and primary place among the substance of war and civilization. Just as group behavior would come to be seen as having its own innate characteristics, irreducible to a collection of individual behaviors, so would situations of man-machine interaction—as between pilot and plane—come to be analyzed as a system, with the specific machinic or human inputs to that system seen as mostly impractical for isolated study.

The abandonment of the model of the disciplined individual as the locus of control required as compensation the adoption of new forms of control, conceived along more decentralized lines. In the control regime initiated by psychologists of the U.S. military, one of the first and most significant tools of decentralized control was the general aptitude test. The modality of the test would come to be adopted at all points where the formal, discreet, and sovereign qualities of the individual had once held sway: in the formation of groups, just as in the ever-more-tightly calibrated feedback loops between man and machine. Hence, it is in the privileged site of man-machine interaction—the cockpit—that we find the prototype of postwar space.

The social sciences formally mobilized for war at a meeting at Harvard University on April 6, 1917—the day the U.S. entered World

War I. A group of psychologists under the direction of Robert M. Yerkes, then professor at Harvard and president of the American Psychological Association (APA), joined the Sanitary Corps at the rank of major to take on the problem of classifying enlisted men. Several days later, the APA offered to "render to the Government of the United States all possible assistance in connection with psychological problems arising in the military emergency."[6] These problems were characteristically manifold. Yerkes lists thirteen general subject areas around which committees of psychologists were formed: aviation; special aptitudes; recreation; vision; military training and discipline; incapacity; emotional stability, fear, and self-control; propaganda; acoustic problems; tests for deception; and adaptation of psychological instruction to military education needs.[7] The emphasis in this list on cognitive skills and training reflected the new realities of modern warfare. For Yerkes, who had previously specialized in animal behavior and was credited, along with psychologist Sergius Morgulis, with bringing Ivan Petrovich Pavlov's work to the attention of North American psychologists, success in war would depend upon the military's use of its cognitive resources, especially in the training and placement of personnel: "Never before in the history of civilization was brain, as contrasted with brawn, so important; never before, the proper placement and utilization of brain power so essential to success."[8]

McDonnell Douglas F-15 cockpit, circa 1984

For Columbia psychologist Edward L. Thorndike, the work done by psychologists during World War I could be roughly divided between "*mass* work," pertaining to the organization of personnel, and "*analytic* work," involving man-machine interactions.[9] Initially, ordering the vast inflow of men into appropriate positions within the

Armed Forces was the most pressing concern. Thorndike cited examples of personnel requisitions filled by the Army's Committee on Classification of Personnel:

> –one hundred and five artists, scene painters, architects, etc., for camouflage work for the Engineer Corps
> –three thousand typists, needed at once
> –forty-five enlisted men capable of leadership who are competent in the distribution and handling of oils and gasolines, fit to receive commissions in the Quartermaster Corps
> –professors of mathematics equipped to teach in the Field Artillery schools
> –meteorologists and physicists able to learn quickly to make meteorological observations and predictions
> –six hundred chauffeurs who speak French
> electric crane operators[10]

Thorndike explained, "In August 1918, nearly four hundred such requisitions calling for over two hundred thousand men were filled. They had to be filled promptly in almost every case, and each had to be filled so as to leave the best possible material to fill every other requisition."[11]

The complexity of these logistical demands represented a true military crisis: in the words of one psychologist, "vague memories of so-and-so's personality and qualifications broke down utterly as a means of building up an army."[12] It brought into sharp focus the military's immediate need to reorganize its handling of personnel. The solution would be found in a newly conceived "personnel bureau," which would seek to recast in systematic terms not only the talents and skills of individuals but also the very nature of "vocation" itself.

Of the tools used by the personnel bureau to match individuals with positions, one of the most essential—and certainly the most debated—was the intelligence test. Work on the general intelligence test had been done before the war, beginning at the turn of the century with the pioneering work of French psychologist Alfred Binet at La Salpêtrière hospital and in the primary schools of Paris.[13] Yet whereas Binet's work was based upon about two hundred school children between the ages of three and fifteen in Paris,[14] the war would provide a venue for intelligence testing on an unprecedented scale. A total of 1,726,966 men, including 41,000 officers, took the Army tests by the end of World War 1.[15] American psychologists developed two "group examinations"—the Army Alpha (for literates) and the Army Beta (for non-literates and non-English speakers)—each of which would require about fifty minutes to complete. The tests were graded on a point scale developed by Yerkes, yielding letter grades ranging from A to E, with A representing high officer potential and E representing unsuitability for armed service. The results were analyzed for their correspondence with the subsequent performance of soldiers in the field.

The Army Alpha included eight sections: the directions or commands test, arithmetical problems, practical judgment, synonym-antonym, disarranged sentences, number series completion, analogies, and information. The Army Beta test was divided into seven sections: maze, cube analysis, X-O series, digit-symbol, number checking, pictoral completion, and geometrical construction. Upon being recruited, a U.S. soldier would find himself sorted into the alpha or beta group:

> A group of draftees, the size of which is determined by the seating capacity of the examination room (it varies from one hundred to five hundred men) is reported to the psychological

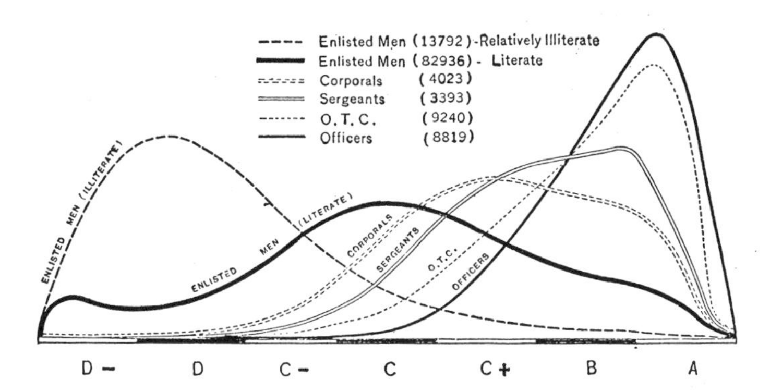

Graph plotting the distribution of intelligence ratings among typical Army groups during World War I, 1920

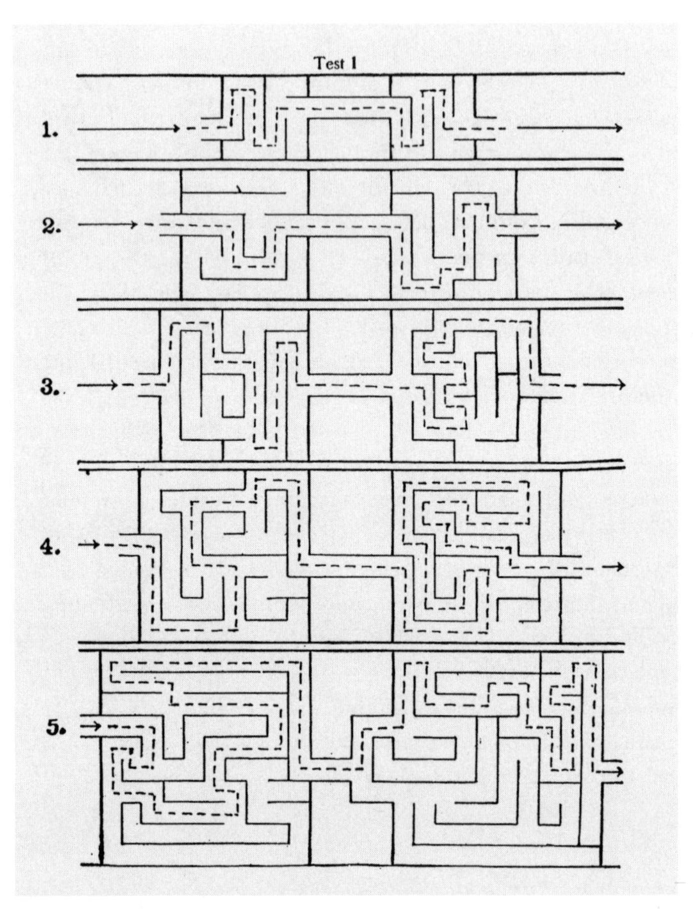

The "Maze," one of a battery of Army Beta group tests, with solutions provided. Prior to each test, examinees were shown by pantomime the solution to a sample problem.

examination room for mental test[ing]. The first essential step is the segregation of illiterates. This is accomplished by having all men who cannot read and write their own letters and those who have not proceeded beyond the fifth grade in school step out of the original group. The remaining men are sent to the alpha room. Naturally, among them there are likely to be several who will subsequently have to

> take the beta examination. The illiterates are sent directly to the beta room.[16]

If a recruit failed these examinations, "either by reason of linguistic handicap or accidents incident to group examining," he would be given a more detailed "individual examination"—either the Point Scale examination, the Stanford-Binet examination, or the Performance Scale examination:

> Every soldier is required to take at least one examination. Men who are of low mentality, those who are of foreign birth or for other reasons illiterate, and those who exhibit marked peculiarities of behavior may be required to take either two or three examinations before their psychological report can be completed. [17]

These tests became the sieves through which passed the entire personnel of the U.S. war machine. The success of the Army tests was such that an official inquiry in November and December of 1917 found that 75 percent of the Army officers acquainted with the tests favored their continuation.[18] For Thorndike, these statistics begged the question of whether the test should be extended to civilian life: "If for the sake of war we can measure roughly the intelligence of a third of a million soldiers a month, and find it profitable to do so, can we not each year measure the intelligence of every child coming ten years of age, and will that not be still more profitable?"[19] If intelligence testing had begun with children, it would return to children on a mass scale. Lewis M. Terman, who held the rank of major during the war, argued,

> Without a knowledge of the quality of its raw material a school has to work more or less in the dark.... Over and over, both in high school and college I have seen students of "A" intelligence, but poor performance, stimulated as a result of the test to improve their work. On the other hand, it is by no means uncommon for an unsuccessful student to be blamed by his instructor, when in fact the cause of his failure is inferior ability, rather than lack of effort.... Men will be found attempting work in lines for which their native ability does

> not fit them. Such students will be everlastingly benefited by being guided into other lines more clearly commensurate with their ability.[20]

Following Terman's recommendation, the Division of Tests of the Committee of Education and Training, in conjunction with the Student's Army Training Corps, went ahead with its program to offer the Army tests to college students, even after the armistice had been signed. Subsequently, the tests were given at 209 colleges, from the Atlanta Southern Dental College, to Brown, Dartmouth, Notre Dame, and the Citadel, to the universities of Texas and California. Test scores were then analyzed according to college, sex, race, field of study, and class year.[21] At the same time, Terman and Yerkes, with support from a Rockefeller Grant, developed from the Army Alpha test a standardized group examination for school grades 3 through 8, called the National Intelligence Test. Completed in 1920, 400,000 of these tests would be available to American public elementary schools in the following year.[22]

For Terman, the benefits of applying the intelligence test across the whole of society were obvious:

> When we have learned the lessons which intelligence tests have to teach, we shall no longer blame mentally defective workmen for their industrial inefficiency, punish weak-minded children because of their inability to learn, or imprison and hang mentally defective criminals because they lacked the intelligence to appreciate the ordinary codes of social conduct.[23]

However, the civilian application of aptitude tests after World War I provoked widespread public debate as to the nature of the "intelligence" measured and the manner in which the results should be used. In 1922 prominent journalist Walter Lippmann published a six-part series in the *New Republic* that criticized what he saw as the technocratic and even anti-democratic implications of ascribing to a coterie of scientists the power of legislating intelligence:

> No student of human motives will believe that this revival of predestination is due to a purely statistical illusion. He will

> say with Nietzsche that "every impulse is imperious, and, as such, attempts to philosophize." And so behind the will to believe he will expect to find some manifestation of the will to power. He will not have to read far in the literature of mental testing to discover it. He will soon see that the intelligence test is being sold to the public on the basis of the claim that it is a device which will measure pure intelligence, whatever that may be, as distinguished from knowledge and acquired skill.
>
> The advertisement is impressive. If it were true, the emotional and the worldly satisfactions in store for the intelligence tester would be very great. If he were really measuring intelligence, and if intelligence were a fixed hereditary quantity, it would be for him to say not only where to place each child in school, but also which children should go to high school, which to college, which into the professions, which into the manual trades and common labor. If the tester could make good his claim, he would soon occupy a position of power which no intellectual has held since the collapse of theocracy.[24]

While Lippmann conceded that "the intelligence test may turn out to be an excellent device for grading children in school" and that it could potentially "grade the pupils more accurately than do the traditional school examinations,"[25] he also saw in the tests the potential dangers inherent in bureaucratic social control, a theme that would later be treated in Aldous Huxley's 1932 *A Brave New World.* The book depicts a nightmare technocracy in which the Army mental test rankings of A through E have been replaced by a genetically and behaviorally conditioned caste system of Alphas, Betas, Gammas, Deltas, and Epsilons, presided over by the superior Alphas, just as the Armed Services, and, later, society itself, would be run by the new "A" men identified by psychology.

The meaning of "intelligence" would also be debated among psychologists, both in pragmatic[26] and ideological terms.[27] Yet we would be misjudging the intelligence test to view it as a mere legislative device that derives its social control through the classification, and hence normalization, of individuals. The "intelligence test" is inseparable from its systematic use within the construct of the personnel bureau, so that the results of testing can only be viewed within the larger context of organizational behavior. Thus the "qualification card" imagined by Yoakum in 1919 as containing "the essentials of the inventory of human material" (including personal history, education, testing data, etc.) cannot be understood without consideration of its function within a personnel bureau simultaneously engaged in the "collection" and "classification of vocational information" (where "each vocation [is] carefully studied from the point of view of the range of mental capacity that will stand a satisfactory chance for success").[28] The qualification card provides an operative, rather than exhaustive, definition of individual qualifications. Its contents are meaningful only to the extent that they are compatible with both a specific systematized body of knowledge on human performance and a specific system of information retrieval.

For an organization, intelligence data becomes a means of access to the skill-set of an individual. Intelligence testing is effective only to the extent that it makes available talent latent in the individual for organizational use. It is not in the interest of an organization to arrest individual development; this can only be viewed as the squandering of a precious resource. The extent to which personnel classification does limit individual development represents inefficiency in organizational operation. As in the design of engines, identifying inefficiencies points toward possible functional improvements. Inefficiency in personnel use demands modification in the organization's process of harvesting individual talent, whether at the level of the classification system or within the organizational structure itself. Of the evolutionary forces that shape organizational structure, one of the most powerful is the need to eliminate such friction in the yoking of individual talent and initiative to organizational ends.

Friction exists within all articulations of organizational structure and at every point of contact between organizational aims and means, whether embodied in technology or individual talent. Like organisms competing for survival, organizations exist in an agonistic environment. For psychologist Raymond Dodge, who taught at Wesleyan and Yale universities and who served in the Navy at the rank of Lieutenant Commander in World War I, the further development of organizational theory had become a military and even economic necessity:

> The military danger is that with the passing of the military crisis we shall stop our study of mental factors in war. If some other country with more permanent policies should take up the mental analyses where we have left them, and develop a real military psychology, they would have a military instrument vastly more effective than 42-cm guns.... [Even in the absence of war,] when mental weapons become the only legitimate means for securing national ends they will become increasingly more important.[29]

During the course of World War I, these "mental weapons" included, in addition to what Thorndike called "mass work," "analytic work" that addressed such problems as determining aptitudes for technical tasks like combat aviation and naval gunnery. Pursued by psychologists such as Dodge and by Knight Dunlap, who served during the war on the Medical Research Board (a branch of the Air Medical Service) and would go on to teach at Johns Hopkins and UCLA, this work applied the modality of the test to the problem of man-machine interaction. Dodge designed what was essentially the first training simulator in the course of developing an apparatus to test naval gun pointers. Thorndike described this apparatus in 1919:

> [Dodge] studied the task of the gun-trainer and pointer, the situations and responses involved, the methods of testing their ability then in use, the men from whom the selections would be made, and the practical conditions which any system of selection for this work must meet. He had the problem of imitating the apparent movements of the target which are caused by the rolling and pitching of the gun-platforms as a distant object would appear to a gun-pointer on a destroyer, a battleship, or an armed merchantman. He solved

> this by moving the imitation target through an 84-phase series of combined sine curves at variable speeds, by a simple set of eccentrics, motor run. He had the problem of imitating the essentials of the control of the gun by the gun-pointer and of recording in a fuller and more convenient form the exact nature of the gunner's reactions in picking up the target, in getting on the bulls-eye, in keeping on, in firing when he was on, and in following through. He solved these by a simple graphic record showing all these reactions on a single line that could be accurately measured, or roughly estimated.[30]

Not only did Dodge discover significant results with regard to the success of the gun-pointer test in predicting future success in the battlefield, but he also found a learning curve among those being tested, with repeated use of the testing apparatus actually yielding improved performance. Dodge had essentially isolated, abstracted, and reproduced a critical interaction between human operator and naval gun interface, including not only gunnery controls but also the rolling movement of a gunnery platform at sea and the evasive maneuvers of the enemy target. Within the simulation, the "intelligence" of the modeled environment, with its programmed time delays representing the behavior of the naval gun platform and the target distances, would interact with the visualization and reaction times of the operator, forging a continuous circuit between man and machine. Within this circuit, both operator and machine simultaneously register and process the behavior of the other, while adjusting their own subsequent reactions. The system records its progress, yielding a simple graphic record of continuously adjusting human response to a responsive mechanical environment and, ultimately, a test score. By the end of the war, the demand was such that the Navy had ordered sixty of these simulators for use in training.

Dodge also developed a number of performance tests and physiological measures for the analysis of the standard gas mask, measuring the speed of mental tasks, visual acuity, eye reactions, eye movements, and finger coordination, as well as degree of comfort while wearing the mask. From this work, "a procedure for determining the effects of different types of masks on the efficiency of the wearer" would be produced.[31] As in the case of the simulator, this line of work required a simultaneous consideration of human factors and machine function. Together, these would constitute a *system*, a man-machine interaction that would effectively replace human subjective experience as the subject of psychology.

The body would become mechanized, or rather systematized—visible only in its reactions with and against a technologically mediated environment, as in the study of the "psycho-physical problems" of high-altitude aviation. In describing his work on high-altitude asphyxia, Dunlap wrote, "The primary effects, we found, are not on any special mechanism or division of the nervous system... but are upon the *integration* of the system, and are evidenced in the decrease in sensory-motor coordination, and in range and sustention of attention."[32] The high altitude, the physical and informational environment of the cockpit, and the respiratory and central nervous systems of the pilot all combine to modulate the critical attribute of attention, as gauged by Dunlap through the measurement of diplopia (double vision) and nystagmus (involuntary eye movements). The system of pilot/plane depends upon this attribute, from the "attention peaks" in which a pilot near unconsciousness or ineffectivity is able to marshal through stimulus and training "brief spurts of normal attention and coordination" in order to retain control of an aircraft during a high-altitude maneuver, to the prevention of "staleness," or "the condition of an aviator when he is markedly unfit to fly for psychological causes."[33]

The work begun by Dodge and Dunlap would come together after World War I under the auspices of aviation medicine and cockpit instrumentation design. Centered at institutions such as the School of Aviation Medicine at Randolph Air Force Base outside of San Antonio, Texas, and the Aero Medical Research Unit of the Air Corps Materiel Division at Wright Field in Dayton, Ohio, the study of "aviation psychology" would address the problem of maintaining pilot comfort and life support in combat situations and at high altitudes. The study of cabin pressurization, rapid decompression, oxygen supply systems, and life support equipment would influence aircraft design,[34] and collaboration between pilots and engineers would lead to the development of the instruments and techniques necessary for precise aerial navigation. The work carried out at these bases would not only influence World War II aircraft design, clearing the way for the adoption of new military strategies based on leveraging control of the air, but it would also lay the groundwork for present-day commercial aviation. The history of aerial navigation illustrates this point.

While the aerial bombing of civilian and military targets had little practical effect on the outcome of World War I, the potential threat represented by strategic bombing was immediately clear. On July 7, 1917, for example, British Air Board officials and civilians alike looked on helplessly while twenty German bombers flew over London, searching unsuccessfully for military targets before eventually dropping their bombs randomly on the city. By the end of the war, it was generally acknowledged that "air superiority" would come to play an important, if not dominant, role in the future of warfare. Airpower advocates such as Italian General Giulio Douhet, British General Hugh Trenchard, and American General Billy Mitchell began to theorize war in three dimensions. Douhet's classic 1923 *Command of the Air* shifted the focus of war decisively away from land and sea battles toward the strategic and overwhelming destruction of civilian and industrial centers of production by airborne forces.[35]

The open-air cockpits of World War I aircraft contained little or no instrumentation. The pilot was expected to rely almost solely on unmediated visual data and "natural instinct" for navigation, landing, and target sighting. Without a "dashboard" or instrumentation panel, the pilot was faced with "look[ing] down at his feet when he wanted to read his instruments or map," rendering his use of navigational tools "a fine art."[36] In addition, the speed of flight greatly complicated the use of existing instruments and navigational techniques, developed as they were for the less time-intensive requirements of sea travel. As celestial positioning was often too time consuming to be practical, World War I-era aerial navigation had to make do with an often inaccurate combination of dead reckoning (estimating one's position using log entries, compass, map, etc., in absence of observation) and pilotage (following known landmarks directly observed from the air).[37]

While a number of flight instruments were available at the end of the war—including the altimeter, airspeed indicator, hand bearing compass, drift sight, and course and direction calculator as well as oil pressure and fuel gauges[38]—neither the instruments nor the techniques for using them were sufficiently advanced for a pilot to fly even a short period of time without recourse to visual landmarks. An accurate turn indicator, necessary as compasses "generally oscillated wildly in turbulence and during accelerations,"[39] had not yet been produced, although a number of possible solutions were being tested.

Naturally, darkness, fog, clouds, and inclement weather posed serious problems for flight. An aerial navigation manual released in 1919 by the Air Service Engineering School and based on the methods developed by the British during the war offered revealing instructions for flying through clouds:

> 1. Select the course to be followed before the clouds are entered and hold the craft to this straight and steady long enough for the [compass] needle to settle to a steady bearing.... 2. When the cloud is entered the pilot must fix most of his attention on the compass and never let the card deviate by more than 5 degrees from the true course, correcting instantly any deviation by a sharp kick on the rudder, using no bank.... 3. Next to the compass, the pilot should watch the airspeed meter and keep the airspeed constant.[40]

Many pilots actually mistrusted flight instruments, preferring to fly by instinct in the case of discrepancies between instrument readings and their own subjective perception of spatial position.[41] Air Force Lieutenant John A. Macready wrote of the first non-stop flight across the U.S.,

> Few people realize that flying is impossible unless there is some exterior, fixed point that the pilot may use to obtain a

> sense of balance or position. If there is no horizon, no light nor any fixed object, a pilot cannot tell the position that the plane is in except from the instruments in the cockpit. When the light disappears a pilot can fly by instruments for a certain length of time, probably fifteen minutes, but he would become very apt to become confused and lose his sense of balance entirely if there were no fixed point within twenty minutes.[42]

The hazards once associated with "blind flying" would soon yield to the certainties of "instrument flying," an invention that synthesized applied research in instrumentation with the training of pilots to rely on those instruments in the absence of visual contact with the earth or horizon. Work on "instrument flying" began in 1918 with a report by William C. Ocker and W. R. Schroeder explaining the potential uses of the gyro turn indicator, a device that was invented and submitted to the Air Service by Elmer Sperry.[43] A pilot, Ocker would emerge as perhaps the strongest proponent of instrument flying. He tested a number of different instrument combinations in the cockpit of a DH-4 aircraft, including one of the first bank and turn indicators ever made.[44]

A critical breakthrough in instrument flying occurred in 1926 during a routine physical examination, as flight surgeon David A. Myers demonstrated to Ocker how a sequence of turns in a Barany chair—a specially designed rotating chair used to test vertigo and spatial orientation—caused every normal pilot to lose track of both the direction and rate of the turns when their eyes were closed. For Ocker, this test demonstrated "the basic reason for the difficulties of blind flight."[45] Soon after, he returned to the examination room with a box jury-rigged to fit around the rotating chair and containing a compass, turn indicator, and flashlight. Using this apparatus, Ocker was able to accurately describe the entire sequence of turns. The breakthrough was the realization that vision alone, and not the subjective experience of spatial position, was the key to blind flight. For Myers, writing later in the *Army Medical Bulletin*, this was "the human standpoint on which the art of blind flying is founded."[46]

A number of pilots, including Ocker, had experimented with blind flight, at Wright Field, Brooks Field in San Antonio, Texas,

DH-4 aircraft instrument panel, circa 1920

and Crissy Field in San Francisco, California. The insights they gained, however, remained mostly at the intuitive level, without a "prescribed course of training" or a "methodology of practicing the skill."[47] The theoretical success of purely visual instrumental flight allowed the focus of experimental blind flight to shift toward the development of methodologies that sought out in a systematic way the best combinations and uses of existing or potential flight instruments. A series of innovations would quickly follow. Modifications of the radio range beacon to produce a homing beacon, along with the invention of Sperry's Artificial Horizon, which indicated the attitude of the airplane above the earth's surface, allowed pilot James H. Doolittle to make the first completely blind take off and landing on September 24, 1929. In 1932 Ocker and engineer Carl J. Crane published the first book on the subject, *Blind Flight in Theory and Practice*, which included three different systems for instrument flight.[48] By 1934 a viable simulator of instrument flight—featuring a radio range-finding device, movable control surfaces, a 360-degree turning radius, and a pneumatic system to simulate the aerodynamics of flight—was in use at six U.S. air bases.[49]

The growing complexity of flight instrumentation would lead to the systematic design of the cockpit instrument panel. For combat aircraft in particular, the flow of information within the cockpit is

Link Instrument Trainer, circa 1934

a critical factor in a pilot's decision-making process. Yet at the same time that the cockpit-as-interface served to mediate the information flow between pilot and plane, the cockpit-as-environment would have to address the physiological and psychological needs of pilot and aircrew. Adding to the complexity of instrumentation as an obstacle to pilot and crew performance would be the aircraft's cramped quarters, the noise of the engines, the cold temperatures and reduction in atmospheric pressure at high altitude, and even the stress of combat. The cockpit would thus become a site of ever-more-careful calibrations of the flight environment to the human body and psyche, and vice versa. These calibrations, aided by military psychology and aviation medicine, would take place both inside the cockpit and within the organization of the military itself. For David N. W. Grant, an air surgeon and brigadier general, speaking in 1943 on the maintenance of pilot effectiveness in war, "the solution is to select and train the individuals best fitted for this duty and to then provide them with devices, methods, and training to assist them and protect them against their limitations."[50]

Psychologists were well enough positioned by their success in World War I to shape the U.S. war machine from the very start of World War II. Thus the single largest program of the second world war, the Army Air Forces (AAF), "was organized under medical auspices" by the AAF Aviation Psychology Program.[51] This grew to include 200 officers, 750 enlisted men, and 500 civilians, and was headed by John C. Flanagan, who was recruited into the military from his former position as associate director of the Cooperative Test Service of the American Council on Education. Under his direction, two widely successful examinations were designed: the AAF Qualifying Examination, a screening exam taken by over one million aircrew applicants; and the Air-Crew Classification Test Battery, which consisted of twenty tests, including six apparatus tests measuring coordination and speed of decision-making and fourteen printed tests measuring intellectual aptitude, perception and visualization, and temperament and motivation. This latter test, taken by more than 600,000 men, was recorded using a nine-category standard score called a *stanine*, which scored individual test performance in relation to the statistical mean and standard deviations of all others taking the test. These scores were found to be highly successful in predicting the success of pilots and navigators in training and combat situations.[52]

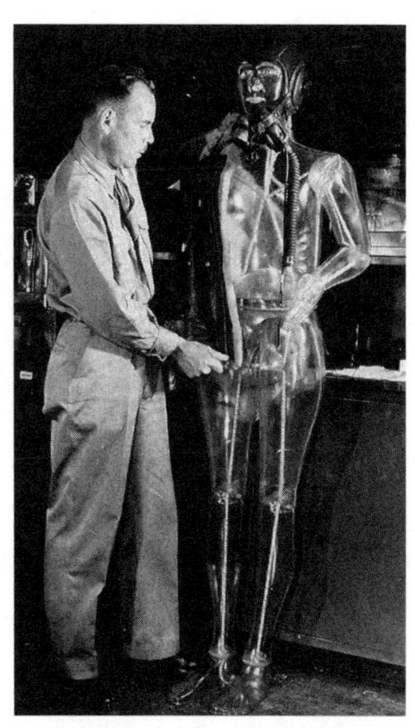

"Plastic Man," built at Wright Field during World War II to test personnel flight equipment and cockpit layout design

Along with this work on personnel selection, World War II also saw the emergence out of psychology of the disciplines of human factors engineering and engineering psychology, which took as their central problem the design of equipment and technology for human use.[53] Following the war, growth in engineering psychology was "explosive," as the time constraints that made prolonged research into equipment design and manufacture untenable were relaxed.[54] Even as the military demobilized, engineering psychology groups increased in number and sought out partnerships with academic institutions and with industry.[55] Although much of this work was driven by the aerospace field—with some of the first successful applications taking place in cockpit design, fire-control systems, and sensor systems—engineering psychology went on to play a large role in a number of different areas, affecting everything from the design of highway systems to the organization of work spaces.

More important than the specific developments mentioned above is the "fusion of the social sciences" identified by Lewin. For the psychologist, this occurred through a marriage of experimental procedure with mathematical analytic tools drawn specifically from economics:

> The analytical tools of mathematical economics should be of great help for carrying through the task of measuring social forces, a task which thus far has been accomplished only in a limited area of individual psychology [namely, the study of decision making]. This task implies three steps; a sufficient development of analytical concepts and theories concerning social forces, their quantification in principle through equations, and measuring concrete cases.[56]

In this narrowing gap between mathematical model and empirical behavior, with each having a reciprocal effect on the other, a new form of social control emerged, perhaps best described through the model of the *simulation*, or *game*. As the methods of empirical measurement grow more precise, and as surfaces, objects, and spaces are designed to more closely fit human use, the feedback loop between human beings and technologies (and also between groups and systems) grows ever more intimate. In addition, as the language of systems became at once more abstract and generalizable, and more specifically applicable to problems in the world, those processes once deemed irretrievably embedded within natural, social, or material interactions became accessible to a rationalization that seeks not so much to denature them as to reanimate them within a controlled environment. In this way the massive wartime reorganization of manpower and materiel persists within our contemporary networked globalism.

The advent of systemization has not been without its effects upon the disciplines of architecture and design. The application of wartime organizational logics transformed both the processes of design and the operative social functions of form. Design—conceived broadly here as that comprehensive set of techniques available to society in defining material problems and their solutions, a social *will to design*, perhaps, that seeks to impress upon the material world the culturally specific attributes of comprehension and utility—no

Robert Probst, at work on an Action Office prototype

longer wishes to limit itself to forms that, once created, cannot be modified to fit their context and environment. Where form implies fixity, design—whether knowingly or unknowingly—after World War II sought to systematically organize change. Accordingly, if we are to make a meaningful connection between the discipline of design and the productive processes that are shaping the world today, we must look toward the design of systems as well as the changing role of the designer in the productive process as a whole, rather than simply the design of form.

After 1945 wide-ranging changes were effected, both in the aims of design and in the methods by which social practices become encoded within spatial locales. No single line of development holds within itself the connate problematics and reconciliations of these changes as fully as that of the office. Indeed, the office has become the premiere site for the systematic accommodation of technological change within a spatial environment. The design of the work environment is often cited as beginning with the 1911 publication of Frederick Taylor's *The Principles of Scientific Management*, with the rationalization of work and the workplace subsequently taking place throughout society as a direct fulfillment of Taylor's program by means of an expanded application of techniques of measurement and a standardization of parts and processes. What was formerly craft production became a "science" through the application of "rigid rules for each motion of every man, and the perfection and standardization of all implements and working conditions."[57] The postwar office environment, however, differs qualitatively from Taylorist production in including environmental change as a formative and primary input; the office became, as in designer Robert Probst's 1968 treatise, "a facility based on change."[58]

Best known for his design of Herman Miller's Action Office (1964), the genetic point of origin of office systems furniture, Probst clearly described how the demand of constant environmental change privileges the design of systems over the design of the formal object. For Probst, the office was a site of "exponential change rates"[59] occasioned by the generation of information within the office environment as well as by the proliferation of new occupations within the growing postwar information economy. Citing Bush's observation that "the scientific community is bogged down in its own product information," Probst saw the office as experiencing the same kind of logistical crisis.[60] Just as technological advances in twentieth-century warfare had demanded a simultaneous decentralization in command structures and specialization in military skill sets, so too would an information economy demand greater organizational flexibility in the mobilization of its technological and human resources. As a result, individual decision making would become ever more crucial to the life of institutions, which accordingly became more flexible with respect to individuals and the tasks they performed. With regard to this development, Probst cites the influential work of organizational psychologist Douglas McGregor, who named the new organizational strategy brought about by the information economy "Theory Y." While the previous organizational strategy associated with mass-production economy—"Theory X"—followed Taylor in granting to management the role of ordering and directing workers who would otherwise avoid responsibility, Theory Y assumed that workers enjoy responsibility and can be counted on to act as managers

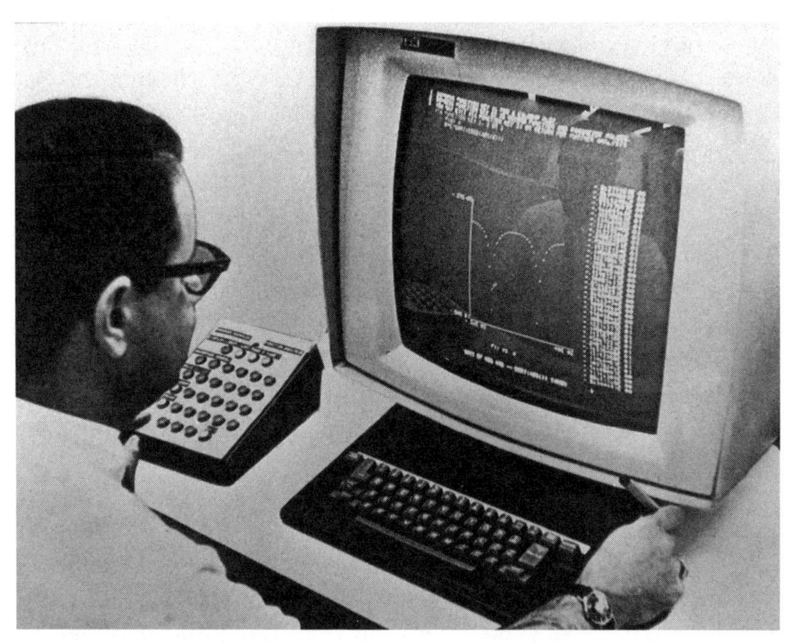

Customized Action Office CAD system

at any level of the organizational hierarchy.[61] It also assumed that both "the capacity to exercise a relatively high degree of imagination, ingenuity, and creativity in the solution of organizational problems is widely, not narrowly distributed in the population" and that "under the conditions of modern industrial life, the intellectual potentialities of the average human being are only partially realized."[62] The challenge, then, would be to draw out these hidden reserves of human potential, so that the "self-actualization" of the worker, to borrow Abraham Maslow's formulation,[63] would be yoked as much as possible to institutional ends.

Probst's Action Office sought explicitly to facilitate a new relationship between institutional hierarchy and its spatial expression. The various functional requirements of the office were met through a modular component system that made use of several types of work surfaces and low wall elements called "standeres." These components could be arranged in a number of ways, both within and outside of a grid system. They could be quickly reconfigured to allow for various kinds of office work as well as various degrees of enclosure and privacy, depending on the communications requirements of the project at hand. Most critical for Probst, however, was the idea that the institution, not the designer, could be in control of the design, the facilities manager working directly on the office floor to determine how the components of the Action Office were to be assembled. Probst imagined the use of modeling and what he called "pre-gaming," or "facility simulation,"[64] in order to virtually enact possible scenarios within the office, using both physical models and the new technology of computer graphics simulation.[65] For Probst, the formal design of furniture was secondary to the design of the systems in which it was used. As in the design of a game, Probst developed a set of rules that would allow for a number of potential outcomes, depending on the game play of the facilities manager. This game play would involve interactions across various scales in the office environment: from the individual office worker customizing his or her workstation, to scenario planning at the institutional level, to the microadjustments of the office floor based on the requirements of day-to-day use.

What matters here is not so much the design object as a singular and sovereign object registering particular material qualities and methods of production, or even the disciplining of the office worker through "scientific management," but rather, the system within which both objects and personnel are to be deployed and the rules and effects of this deployment. Here, individuals become active agents within a fluid hierarchy or set of protocols that extract from them not simply the free formal labor elaborated by Karl Marx but the sum of their individual skills and initiative. The subsequent history of the office and the workspace—from telecommuting to lean production—can be seen as further progressive alignments of the goals and self-image of the individual worker with those of the institution itself. The result is a blurring of the formal boundaries between institution and individual, just as human factors and engineering psychology effaced the distinction between human being and machine within the man-machine system. As psychologist

Franklin V. Taylor pointed out in 1957 in a remarkable paper entitled "Psychology and the Design of Machines," this necessitates the abandonment of the anthropocentric model in favor of the "mechanocentric."[66] For Taylor, writing from the U.S. Naval Research Laboratory in Washington, D.C., "man" disappeared as a defining issue in the psychology of machine design, to reappear only as an "operator," a "system component," or an "organic data transmission and processing link between the mechanical or electronic displays and controls of the machine."[67] The essential behavior here occurs at the level of the system. Even within competing systems performing the same basic functions, the "human component" may contribute in very different ways:

> One system may require the operator to act analogously to a complex differential equation-solver, while another may require of him nothing more than proportional responding. One radar warning system may require the operator to calculate the threat of each target and to indicate the most threatening; another may compute the threat automatically and place a marker around the target to be signaled.[68]

"Man" is defined within the system in a language appropriate to that system; whether he appears as "information channel," "multipurpose computer," or "feedback control mechanism" depends on the use value of these models in understanding (mathematically or otherwise) the behavior of the system as a whole.

At the level of the system, the design of the cubicle recalls the design of the cockpit. If the Action Office sought to better integrate office workers and office work flow, then the Kinalog Display System, designed in 1959 by the General Dynamics Corporation for use in combat aircraft, sought to "achieve a maximum compatibility between the pilot and his cockpit," by better integrating the perceptive apparatus and sensori-motor function of a fighter pilot with the performance environment of the combat aircraft.[69] The Kinalog Display System was a new kind of "attitude display," a device that was already a standard part of the cockpit instrumentation panel that informed the pilot of the relationship, or "attitude," of the aircraft vis-à-vis the earth's surface and hence indicated which way was "up." The Kinalog display took the place in the instrumentation panel of the prior attitude display and thus had essentially no formal impact on the panel layout; its impact would be performative, in its behavior as an interface. Whereas earlier attitude displays were either "inside-out," displaying the aircraft as fixed referent and the ground plane as variable, or "outside-in," displaying the ground plane as fixed and the aircraft as variable, the Kinalog display made use of the entire continuum between these two display systems, so that both aircraft and ground plane became variable, with the rate of translation between the two systems dependent on the continuously changing G-force sensed electromechanically within the aircraft. While all three display systems—inside-out, outside-in, and Kinalog—present the same information with regard to the relationship of aircraft to ground plane, the Kinalog is designed to register the subjective experience of G-forces by the pilot and thus "resolve the sensory conflict between visual and kinesthetic inputs,"[70] which can cause pilot vertigo as well as interfere with the pilot's intuitive sense of spatial position. The Kinalog display, and by extension the cockpit instrumentation panel itself, is tuned to human adaptation to changes in gravitational force. While the Kinalog was certainly designed as a concept, its material design and behavior could only have emerged through the combined input of pilot and plane.

Likewise, the Action Office sought to be sensitive to input at various scales within the office organization, so that its layout in a sense would be self-organizing. This self-organization stands in marked contrast to the fixity of architectural form, or even the ostensible formal autonomy of, for example, modernist furniture design. If form is secondary to system in the Action Office, this is because form functions as only one of a series of transactions that animate the Action Office. Such transactions would go well beyond satisfying the basic demands of office functionality. The components of the Action Office would constitute a protocol of user interactivity in defining the various ways a worker might modify his or her workstation over the course of executing daily tasks. In substantiating such a protocol, it would bring into play the system as

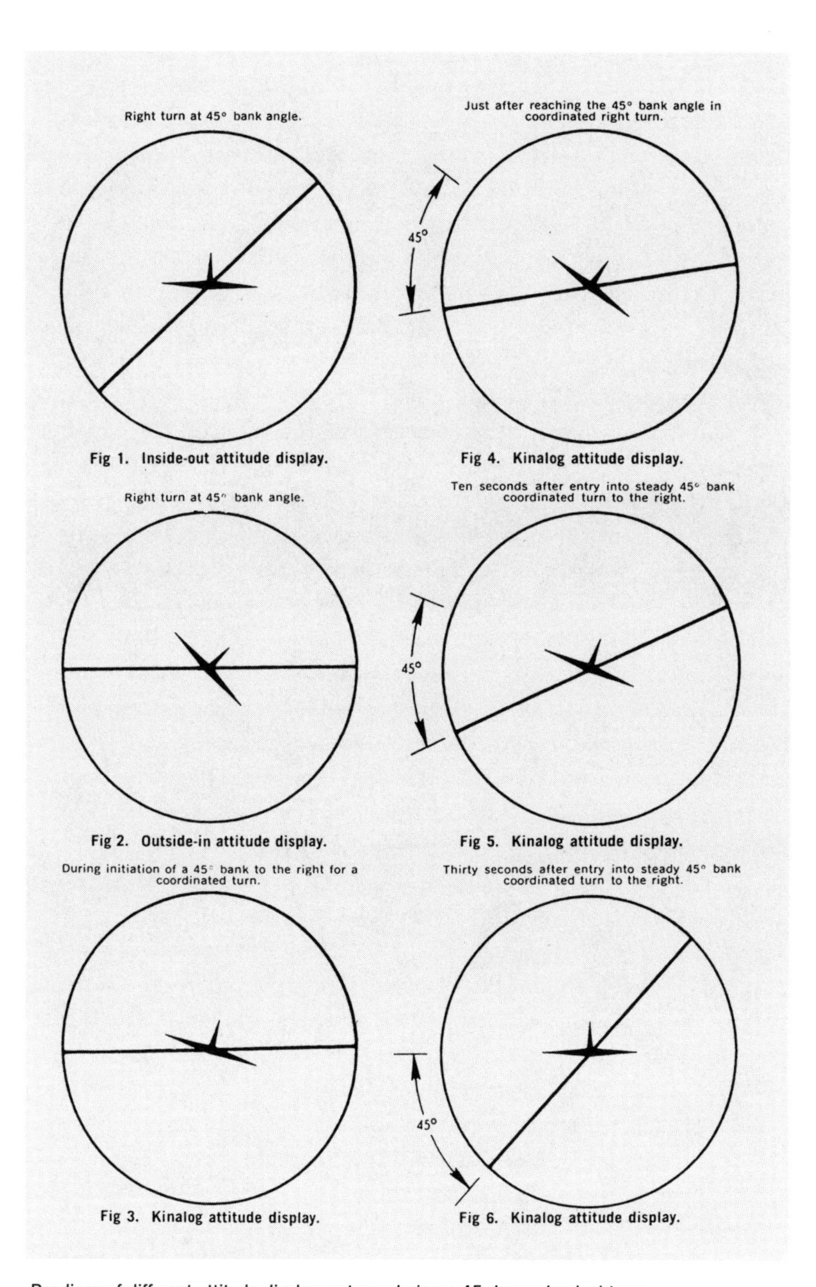

Readings of different attitude display systems during a 45-degree banked turn

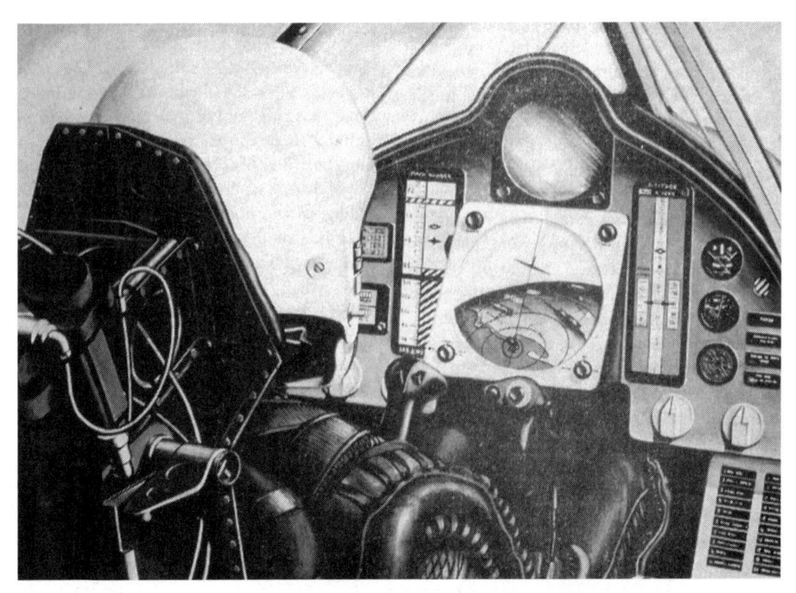

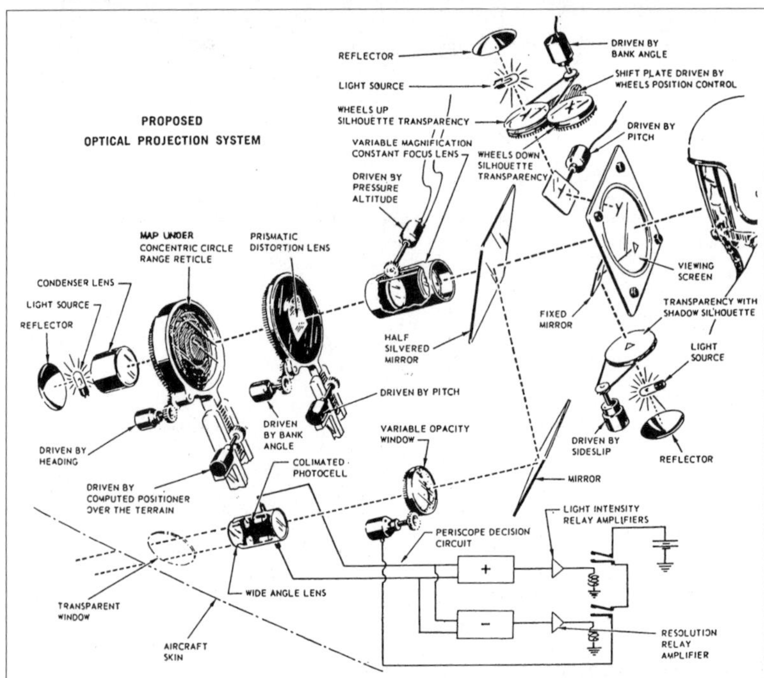

The Kinalog Display System, set into a cockpit instrumentation panel (top); Proposed optical projection diagram for the Kinalog Display System, designed by the General Dynamics Corporation for use in a cockpit instrumentation panel, circa 1959 (bottom)

an idea, both as an ideology that connotes a certain kind of business practice and as a virtual set of relationships to be simulated in the "pre-gaming" of the office layout.

If the central socio-technical dilemma of postwar North America was ultimately that of assimilating a profound shift from mechanocentrism to systemocentrism, this shift would be reflected in a widespread preoccupation across a number of disciplines—among those, most certainly architecture and design—employing nascent information technologies and cybernetics. The gradual waning of interdisciplinary systems thinking in the 1970s, however, should not be seen as signaling the abandonment of the socio-technical systems model. If anything, post–Cold War globalism only serves to emphasize the kinds of interdependencies that define the behavior of systems. While a certain postwar idea of *planning for complexity*, well-represented in such cultural artifacts as the Action Office, did fall out of fashion, one could argue that this was simply the result of the migration of control from the planner to the economy, yielding a market-centrism within which we are still enmeshed. The kinds of power over humanity once attributed to the planner and to the intimate machinic sensitivity to the human body that emerged from the study of ergonomics have only been enhanced by the unparalleled ability of the market to regulate change.

The postwar preoccupation with systems forms an important precursor to the current state of affairs. Within the pandemonium of the global market, it is fruitful to consider the time when it was thought that complexity could be planned. Certainly the proposed mechanisms of the systems era—where the very operation of systems works to make relative the meanings assigned to those forms within their matrices of control—will remind us of the tenuous position of form in culture. Today, the claim to criticality through the destabilization of form (by hybridization, etc.) is essentially problematic as it is precisely this destabilization (as in the destabilization of anthropocentrism by organizational and engineering psychology) that occurs with the operation of system upon form. It is no longer possible to see form as a citadel, self-contained and sovereign, imposing a fixed symbolic or use value upon a previously undifferentiated and hence free environment. The only remaining criteria that now matter are those that are operative, that actively determine the relative positions of elements within their given field of operation. Every element within the domain of the system is rendered active and mobile simply by virtue of its compliant existence as a constituent part of the system as a whole, while the symbolic and use value once resident in form may only be reconstituted within the functional, systematic correlation of these elements. We can now only see the form-meaning relationship as provisional and variable, dependant upon the system within which it is located. In this way, form is both denatured and opened up to a number of potential meanings.

FORECAST

ANNMARIE BRENNAN

Language is the armory of the human mind; and at once contains the trophies of its past, and the weapons of its future conquests.

—*Samuel Taylor Coleridge*

Crawling out of a German U-boat and onto the dark and foggy Long Island coast in the early morning hours of June 14, 1942, eight Nazi agents, with only a suitcase-full of American currency, explosives, and dry civilian clothes, sought to destroy the manufacturing core of the United States war effort. The plan, termed Operation Pastorius, involved the destruction of strategic industrial and logistical sites such as hydroelectric plants at Niagara Falls, a cryolite plant in Philadelphia, the locks on the Ohio River between Louisville and Pittsburgh, the water system for New York City, and the Pennsylvania Railroad Station in Newark, New Jersey. In addition, the Nazi agents were ordered to "harm as much as possible aluminum production in the United States,"[1] targeting the Alcoa aluminum plants in Alcoa, Tennessee; Massena, New York; and St. Louis, Illinois.[2] In the end, the saboteurs safely reached the American beach and were able to board the Long Island Railroad and travel as far as New York City's Pennsylvania Station, where they were eventually captured by the F.B.I. Aware of the strategic importance of aluminum in making American military warplanes, the Nazis sought to destroy the fount of its manufacture. American plants were responsible for over 40 percent of the world's aluminum output in 1943.[3] This massive yield of aluminum would later be reassigned from the battlefield to the domestic realm.

The unprecedented increase in U.S. production capacity during wartime was not the only industrial and technological achievement spurred on by World War II. The implementation of systems analysis and operation research, an innovation developed by the military-industrial-academic think tank RAND, would change the method of modern warfare. Systems analysis was initially applied to conditions in which the stakes were at their highest, where scientists could only theorize the outcome of a war scenario, such as the retaliation of an enemy force using an atomic bomb. The challenge was to design the problem, criteria, objectives, and a series of possible conclusions through simulations, quasi-experimentation, and operation gaming (referred to as RAND games). The processes used in systems analysis would come to be known as "forecasting."[4]

Like aluminum production, technological forecasting techniques developed in World War II to gain territory and bomb cities would be redirected toward improving the postwar American domestic domain, where advertising campaigns based on the same military logic ultimately found their place in the hands of most Americans via the pages of weekly magazines such as *The New Yorker*, *Time*, and *Newsweek*.[5]

The marketplace assimilated military logic and tactics developed through the establishment of the military-industrial complex. The

THE SATURDAY EVENING POST April 28, 1945

To have butter, lots of it, will seem heaven enough.

And then to have it as fresh as if churned the day you buy it! That is how good butter will be when it comes to you wrapped in aluminum foil. (You can even let it get chummy with melons or fish in the refrigerator. The foil will protect it from foreign odors and tastes.)

When Alcoa Aluminum Foil returns from war, you will see all manner of foodstuffs wrapped in it to keep them fresh longer. Not too dry. Not too moist. Natural flavor, color, and goodness sealed in by a friendly metal barrier. Enemies of freshness sealed out.

ALUMINUM COMPANY OF AMERICA, 1843 Gulf Building, Pittsburgh 19, Pennsylvania.

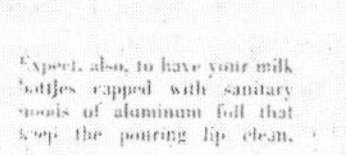

Cheese is another dairy product that stays fresher when wrapped in aluminum foil. Doesn't dry out. Doesn't lose flavor.

ALCOA ALUMINUM FOIL

ıns for butter. Alcoa advertisement, as it appeared in the *Saturday Evening Post*, April 28, 1945

postwar marketing campaign of the American Aluminum Company of America (Alcoa) provides evidence of this trend. FORECAST, as the campaign was known, enlisted architects and designers to create future scenarios with aluminum as its design inspiration. The program provides an interesting model, illustrating the application of forecasting as a design tool as well as demonstrating the need for American manufacturers to adopt design as a strategic selling tool in the postwar market.

"ALUMINUM FOR WAR AND PEACE"

Before the United States entered into World War II, President Franklin D. Roosevelt realized the strategic importance of the airplane to the war effort. On May 16, 1940, he announced in a speech to Congress the government's plans to construct 50,000 planes a year for the next two years. On January 6, 1942, the President called for 60,000 planes to be manufactured that year and 125,000 in 1943. The requirements of aluminum production were inextricably tied to the escalating call for more military aircraft. To meet increased demand for aluminum, U.S. industry increased its production from 206,300 tons in 1940 to 920,000 tons in 1943.[6]

An April 1943 *New York Times Magazine* article "Aluminum for War and Peace," written by journalist R. L. Duffus, described a Tennessee Valley Alcoa aluminum plant during a tour of a rolling mill:

> The great building was shuddering with the impact of metal as aluminum slabs weighing 3,200 pounds were grabbed by the rollers, on their way to become wings for fighting planes. The manager had to shout to make himself heard. "This," he roared, "is an aluminum production war." He was boosting his own product, but he had the figures to prove his case. Without aluminum for planes there would still be a war. But without planes it wouldn't be this kind of war.[7]

The importance of aluminum in military aircraft was its strength in relation to its weight, which is one-third less than steel and other metals.[8] Aluminum, as opposed to an equal amount of steel, allowed a plane's wings to be longer yet thinner, thereby allowing the plane

THE TIMETABLE OF ALUMINUM FOR VICTORY

1938 Sept. *Munich.*
Oct. *Czechoslovakia invaded.*
Nov. Alcoa inaugurated an expansion program which by 1943 cost $250,000,000.
Dec. Alcoa 1938 production, 287 million pounds; started 1939 with more than a year's supply on hand.

1939 Jan. Alcoa begins operation of new extrusion and tube mill.
Feb. Alcoa starts building excess stock pile of airplane sheet.
Apr. *Congress authorizes Army to acquire 6,000 planes in 2 years; and Navy 3,000 in 5 years: Approximately one month's 1943 goals.*
Sept. *Poland invaded.*
Alcoa authorizes new metal-producing capacity.
Nov. *Finland invaded; Cash-and-Carry Act signed.*
Dec. Alcoa authorizes another huge metal-producing plant, although it begins new year with 215 million pounds on hand.

1940 Jan. *First Request for defense appropriation in Budget Message.*
Mar. Alcoa reduces price of ingot from 20c to 19c.
Apr. *Denmark, Norway invaded.*
May *Low countries invaded.*
First of new Alcoa metal-producing plants starts production.
June *Dunkerque; France capitulates.*
Alcoa authorizes still another metal-producing unit.
July *Congress gives first go-ahead on faster plane production.*
Aug. Alcoa reduces ingot price from 19c to 18c; adds large alumina capacity.
Sept. *Egypt invaded; Selective Service Bill passed.*
Another new plant starts operating; still more units authorized.
Oct. *Rumania invaded.*
Nov. Alcoa reduces ingot from 18c to 17c; authorizes still more metal-producing capacity.
Dec. Alcoa faces new year with 154 million pounds on hand.

1941 Jan. *NDAC says aluminum supply adequate to meet October, 1940, estimates of requirements.*
Feb. *Aluminum put on priorities to give all capacity to defense.*
Mar. *Lend-lease.*
Apr. *Yugoslavia invaded; U. S. occupies Greenland.*
May Very large new Alcoa metal-producing units start operation.
June *Crete lost; Russia invaded.*
Alcoa authorizes further expansion at own expense.
July *Government authorizes first of its own plants to supplement the enormous expansion of Alcoa.*
Aug. *Government announces Alcoa will build and operate 3 of these plants. (All Alcoa designing and building of Government plants done without profit.)*
Sept. *Government decides on more plants; instructs Alcoa to build them.*
Oct. *Government decides to build more plants; instructs Alcoa to build them.*
Alcoa reduces ingot from 17c to 15c.
Nov. *Government reviews detailed plans for own large sheet mills.*
Dec. *Pearl Harbor; Churchill-Roosevelt strategy conference.*
First metal rolled on Alcoa's own 50-times-faster sheet mill, largest in the world. Alcoa receives further new plant instructions from Government.

1942 Jan. *Pan-American Conference.*
Another Alcoa metal-producing plant in operation; additional instructions from Government to build new plants.
Feb. *Government authorizes large alumina plant, several aluminum plants and large sheet mills. Alcoa to design and build.*
Mar. *Government authorizes Alcoa to build more casting capacity, extrusion capacity, and forging capacity.*
Apr. *Government authorizes blooming mill, and enlarged tubing capacity.*
May *Government-owned plants, built and operated by Alcoa, start operation.*
June *Special authorization for airplane cylinder-head capacity.*
July to Dec. { *As instructed by Government, Alcoa starts additional plants for various types of fabrication, as special needs of war production are made apparent by changing emphasis on war equipment.*

1943 The aluminum industry will have a metal capacity of over two billion pounds, seven times prewar. Alcoa has more than doubled its metal producing and fabricating capacity through a self-financed expansion program. The expansion by private industry has been augmented by a vast Government program where the kind, amount and time of expansion has been at the direction of the Government. In addition to operating its own twenty plants, Alcoa has been honored with the responsibility for constructing and operating 40 Government projects in 25 different locations.

ALCOA ALUMINUM

"Alcoa Timeline of Aluminum for Victory," *Newsweek*, April 19, 1943

to travel at an accelerated speed and/or farther distance using the same amount of fuel,[9] a crucial factor when under attack by another warplane.

Duffus put the increased production of aluminum in perspective for his readers:

> By the end of this year the United States will be producing at the rate of 2,100,000,000 pounds of aluminum a year....It is sixty-times the production of 1910; fifteen times the production of 1920; nearly ten times that of 1930; about seven times that of 1938; about five times that of 1940; It is half again as much as the world produced prior to the beginning of the present war. It eats up enough electricity to light every home in the United States for nearly two years. It is a hungry metal with a hungry market.[10]

World War II demonstrated to the world the capacity and power of American industrial production and its ability to expand exponentially to meet wartime needs. While the overall production rate of industrial raw materials such as metals and chemicals increased more than 60 percent during the war and the production of steel doubled, the output of aluminum increased to almost 400 percent.[11] This increase in production was the direct result of the massive fabrication of warplanes. Ninety percent of the wings and fuselage, 60 percent of the engine, and all of the propeller was comprised of aluminum—all tolled, 75 percent of an airplane's entire weight. Rivets, wires, cables, rods, radios, instrument cases, cockpit fittings, aerial cameras, and the hydraulic system that opens and closes the bomb doors and retracts the wheels of the plane were also made of aluminum.[12] The ratio of aluminum-to-airplane was a simple one: 1:1. By calculating the weight of an airplane, one could estimate the cost of aluminum and, therefore, the entire cost of the airplane.[13] This equation of aluminum-to-airplane was exploited by Alcoa's advertising agency as a synecdochical relationship, where the material—aluminum—was characterized by the very object that it formed—the heroic and victorious World War II airplane.

"THE ACADEMY OF DEATH AND DESTRUCTION"

In 1942 a group of specialists from the Office of Scientific Research and Development (OSRD) and the Office of the Secretary of War went onto the battlefield with the purpose of educating soldiers about the use of new weapons and military tactics. While in the field, the group conducted investigations and analyzed the success rates of new equipment and techniques. This type of procedure would come to be known as "operations research" (OR).

During World War II, one of the first Americans to bring OR into normal military practice was Dr. Edward L. Bowles. As consultant to the Secretary of War and founder of the Massachusetts Institute of Technology's Radiation Laboratory, Bowles pioneered the development and implementation of radar within a military setting. More importantly, Bowles was one of the first, along with Vannevar Bush, founder and head of the OSRD, to advocate the creation of a nonprofit, civilian advisory group consisting of scientists and industrialists to consult with the military on technological advances and strategic practices.[14] This group would be responsible for OR tasks such as gathering data about "meteorological conditions, combat charts, numbers of rounds fired versus enemy hit... and put[ting] them through the number cruncher to try and improve the effectiveness of both weapons and tactics."[15] OR was also defined as "applied scientific method to the use rather than the technology of weapons."[16]

One of the first attempts to increase the direct participation of civilian consultants within the military began in early 1944 and centered around the B–29 Special Bombardment Project.[17] Working with the Army Air Forces and the War Department, a team consisting of civilian engineers from the Boeing and Douglas Aircraft companies were recruited to study "the effectiveness of the B-29 as a strategic bomber."[18] The group reported that the bomber's performance could be enhanced by dismantling its unnecessary heavy defensive armor. By eliminating extraneous weight, the bomber improved its range and bomb load; it also increased its speed to surpass that of any known Japanese fighter plane.[19]

A. E. Raymond, vice president of engineering at Douglas Aircraft, understood the advantages of having a scientific organization research to "improve foresight in aircraft requirements and give the aircraft industry the means to reduce waste in design."[20] This effective partnership between scientists, industry, and the military sparked conversations at the War Department and the OSRD. With the end of the war in sight, what would happen to the seemingly short-lived relationship between military planners, civilian scientists, and industry? On November 7, 1944, Air Force General H. H. "Hap" Arnold wrote a memorandum explaining the important role of research and development, urging the Air Force to retain the fruitful relationship begun between scientists and the military after the war to "assist in avoiding future national peril and winning the next war."[21] This memo began what is known today as Project RAND.[22]

Project RAND created an environment in which the government could, after the war, seduce scientists to remain as consults for the military rather than return to their university positions. It was understood that the military needed more than brawn—it required intellect. It was necessary for the government to find some way to retain the scientific workforce that had developed the atom bomb, the radar, the jet plane, and other technologies of modern warfare.

Funded by a research war surplus of $10 million, General Arnold signed a contract with the Douglas Aircraft Company in late 1945.[23] A portion of a Douglas Airplane hanger in Santa Monica, California, was sectioned off, and Project RAND was founded. On March 9, 1946, the program was officially defined by General Curtis E. LeMay, commander of the Strategic Air Command, as "a program of study and research on the broad subject of intercontinental warfare other than surface" with the objective of recommending to the Army Air Forces "preferred techniques and instrumentalities."[24]

LeMay's statement of work was intentionally vague, allowing the project to have a degree of independence in its area of study. Some of the research topics addressed in the early years of Project RAND included rocket engines for strategic weapons, boron and other high-energy fuels, the statistical theory of radar detection, game theory, air defense, nuclear propulsion, metal fatigue, optimal design structures for military aircraft, and bomber and fighter design.[25] In terms of bomber and fighter design, RAND would not create a specific blueprint but would make technical recommendations for new bombers, suggesting specific enhancements to improve missile precision, dependability, effectiveness, and cost. Project scientists would also investigate aspects of strategic planning in an attempt to ascertain the most effective bomb, attack formation, and psychological weapon for a quick surrender.[26] The intentions and capabilities of Soviet Russia were the central focus of RAND's research during the Cold War, and it is perhaps due to this concentrated attention that the Soviets deemed the RAND Corporation the "Academy of Death and Destruction."[27]

Project RAND and its work were kept secret until March 1951, when a *Fortune* magazine article introduced the public to its methods and aims. The article's author, John McDonald, described the initial goal of Project RAND as "improving the research and development decisions that determine the character and set the limits of future combat."

> RAND's scope, therefore, encompasses all conceivable future Air Force weapons and conveyances—some not yet in existence—and the strategies of their use for offense and defense. "Future," for the most part, means at least five years hence; it normally takes that long to get through experimental development and mass production to front-line operation. But while the focus is years ahead, the problems and deadlines are current; for the Air Force of tomorrow is only as good as the decisions it makes today.[28]

The existence of RAND was announced to the public within the pages of a business magazine; the lines between the government, industry, and science were forever blurred, woven together to fabricate a single complex—a huge organization working with similar methodologies toward a common objective. The *Fortune* article compared the technological forecasting performed by RAND with the forecasting that might be done by an industry and maintained,

Alcoa advertisement, as it appeared in *Life* magazine, February 14, 1944. The caption reads, "You never figured, did you, Mister Hitler and Mister Yamamoto, that just this one outfit, sixty thousand Americans with an awful lot of know-how, would be able to push out so much metal to make so many planes so soon? And the thousands are on their way to becoming ten thousands."

"In a sense this is the familiar business problem of providing new facilities and making replacements and other changes for an uncertain future."[29]

The first major project for RAND suggested by the Air Force was the preliminary design of an "experimental world-circling space ship." The study, conducted as early as 1946, researched the feasibility and military advantages of a satellite program. RAND concluded that a satellite could be launched by 1952 and that the scientific data collected and transmitted to Earth would be "extremely valuable." The vehicle would have "important military uses in connection with mapping and reconnaissance, as a communications relay station, and in association with long-range missiles."[30] With the increased importance of military research and development, Project RAND decided that it needed to become a more independent entity, distancing itself from Douglas Aircraft and any feelings of privilege the airplane company felt. In 1948 RAND established an advisory board, with members including Douglas Aircraft, North American Aviation, Boeing Aircraft, and the Northrop Corporation. It also decided to expand its realm of study subjects, conscripting engineers and social scientists to examine systems analysis, psychological warfare, and propaganda.

Operations research and systems analysis depended on the development of the computer, which, unlike humans, was able to store and quickly calculate huge amounts of information. Operations research collected data from past experiences and applied mathematical analysis to it, while systems analysis attempted to *predict* the outcome of specific situations based on mathematical probabilities. A strategically important military instrument for combat situations, the computer was implemented in both operations research and systems analysis, compiling and calculating research data for trajectories of anti-aircraft fire.

Systems analysis and forecasting would come to infiltrate aspects of society requiring long-range planning, from manufacturing, technology, and economics to urban development and weather tracking. Dr. Alian C. Enthoven, an economist working at RAND on applying "economic principles to strategy in the Defense Department" drew a parallel between the application of super-rational theory to economics and thermonuclear warfare:

> By 1961, a great deal of progress had been made in the development of an economic theory for our posture for thermonuclear war.... [W]e have made a great deal of progress in the translation of our broad objectives into specific quantitative criteria that can be applied in a systematic and practical way to the evaluation of proposed forces and postures.... *The economic theory of our posture for nuclear war can be described in terms very similar to the economic theory of a multi-product firm.*[31]

Thus the same strategy applied to thermonuclear war could function as a business plan for a postwar corporation. Thermonuclear warfare and the capitalist marketplace shared the same terminology, and their methodologies were equivalent: capitalism behaved in a manner similar to warfare.

"A MAKER OF METAL"

The immediate problem facing the aluminum industry after the war was not conversion to peacetime needs, nor the disposal of government-owned plants; these issues would be resolved later. The principal dilemma entailed reconversion and surplus—"surplus capacity, surplus stocks, and surplus scrap."[32] A report from the War Production Board stated, "It would seem that after the war either the metal industries will have to shrink to fit the rest of the economy; or the rest of the economy (or exports) will have to expand sufficiently to permit reasonably full utilization of the present facilities and labor force of the metal industries."[33] Would plants be closed, or would there be a way to continue the accelerated rate of production during peacetime? A 1945 study conducted by the U.S. Department of Commerce and the Washington State Planning Council on the postwar outlook for aluminum maintained,

> Looking to the future, the transition from wartime to peacetime uses of aluminum must take place at the fabricating stage, new products must be discovered, engineered, and produced or there will be fewer jobs at all stages of the aluminum industry. Skilled metal workers now in aircraft plants and those now working in the reduction stage of the aluminum industry will have to seek work outside aluminum manufacturing unless new end uses of the metal are developed.[34]

In the aftermath of World War II, the War Production Board ordered the creation of an aluminum stockpile. It did so in order to avoid the chaos and expense of rapidly building manufacturing plants on the "rare" occasion of a future war. After 1948 when the Soviet Union created an atomic bomb, the aluminum industry benefited from continued government investment "in the abnormal context of a defense economy."[35] Rather than shut down many plants, the government kept them open in order to amass this stockpile, assisting the aluminum industry and the general economy along with the military. In the years that followed, the Korean War would also play a crucial role in perpetuating high production capacities.

During 1957 and 1958, Alcoa and other primary producers of aluminum were undergoing aggressive capital expansion programs when an economic recession hit. The recession sent the price of aluminum down, while creating an extended period of excess capacity.[36] The aluminum industry needed to find an answer to the questions, into what new markets can aluminum enter, and how can we increase demand?

The question of increasing demand was initially answered by one of Alcoa's biggest competitors, the Reynolds brothers. Immediately after the war, Reynolds had begun to fabricate finished products such as aluminum toys, rowboats, home freezers, golf clubs, cooking utensils, and other domestic goods. This initiative helped frame aluminum as the quintessential material for new uses in the postwar domestic market. Experienced in distributing and marketing consumer goods, Reynolds, quick to fashion aluminum into finished products, would eventually expand into larger durable goods such as windows, automobiles, trailers, curtain wall systems, and electrical products.

By comparison, Alcoa sold its product—unfinished pig aluminum—to manufacturers and to the government for the manufacture of finished goods. Before the war, Alcoa—the only company producing aluminum on a large scale, hence its title "the Aluminum Company"—dealt mainly in producing and selling crude or semi-crude products, since the front-end of the production cycle was where the deepest profits were to be found. Alcoa was wary of delving into semi-finished products for fear of competing directly with too many of its own customers. Reynolds, a relative newcomer to the aluminum business, was not concerned with established industrial allegiances.[37]

Unlike its competitor, Alcoa was not experienced in producing or marketing "value-added" finished products. It attempted to expand into these markets but found itself competing with many of its own clients. The option of buying a subsidiary to improve its profit margin was dangerous for Alcoa since, given the company's history as a monopoly, the federal government was constantly looking over its shoulder for antitrust violations.[38] Alcoa would have to solve its excess capacity problem another way.

The war had increased the demand for, and therefore the value of, semi-finished and finished products. When military demand lessened in the postwar period, Alcoa began to understand the economic complications of a competitive market. The company was initially slow to act, maintaining the same position in the marketplace as before the war, as "a maker of metal."[39] The company wanted to continue dealing in crude and semi-finished tonnage rather than small, finished aluminum products. Its ideology was based on the idea that with "every ton of aluminum...sold to provide skin for a new office building was a ton more production for Alcoa's smelters."[40] The strategy of concentrating on only large-scale, semi-finished custom projects such as aircraft and machinery became outdated when the price of aluminum began to fall. Alcoa was now required to participate in the entire aluminum production chain and to begin concentrating on the end of the production cycle, applying the value-adding strategies of design and advertising.[41]

THE ROLE OF THE PROPAGANDA PLANNER

Project RAND's area of inquiry was not limited to operations research and systems analysis; the organization also conducted research into psychological warfare. In the study *U.S. Wartime Propaganda: The Role of the Propaganda Planner*, completed in 1950, RAND researchers described the propagandist's tasks as conforming to the following criteria:

> [The tasks] should aim (either directly or indirectly) to induce an audience to perform specific actions, and they should be capable of at least partial accomplishment by the media at the disposal of the propagandist. Certain tasks may be directed toward building audiences and credibility, or toward influencing attitudes and expectations, but these are merely preparatory measures for the tasks which aim at inducing specific actions. Propaganda can influence action by persuading an audience that a given situation will exist in the future, by persuading an audience that a given situation already exists, by deceiving an audience as to the true situation (i.e., persuading it that a given situation exists, when in fact it does not), by encouraging attitudes which are likely to result in action favorable to one's own cause and unfavorable to the enemy, by encouraging non-rational or emotional action, by rewarding individuals for certain past actions, and providing know-how.[42]

Through media such as the Voice of America radio broadcast and U.S.-sponsored magazines, propagandists gave the impression that a certain condition or scenario existed, or would exist, in the future. The RAND report stated, "If people accept this portrait of the future, they are more likely to orient their actions accordingly than they otherwise would be. The most frequent statement about the future made by the [Office of War Information] was, of course, that the U.S. would win the war."[43] The power of these techniques was their ability to obscure the difference between propaganda (where a statement is made to bring about a certain future condition) and forecasting (where a statement is made about a likely future condition). Propaganda methods created and practiced by RAND would eventually become the rhetorical methods applied by companies such as Alcoa to postwar marketing and advertising as a means of increasing sales and profits.

Alcoa's initial strategy in finding new customers did not rely on advertising but on its own Research and Development Department. As a result of Alcoa's reputation as a monopoly, combined with disputes over industry patents developed during the war, the U.S. government and its think tanks were reluctant to work directly with the aluminum company; Alcoa did consult, however, with the OSRD and the Manhattan Project and, throughout the Cold War, would work on the development of alloys for aircraft with hothouse industry insiders such as Boeing, Hamilton Standard, and Douglas.[44]

In addition to these partnerships, Alcoa began to forge alliances with the construction industry—the largest growing sector of the American economy. By 1946 this industry would consume almost 20 percent of the company's total output of aluminum, compared to a prewar figure of 8 percent. The architectural firm Harrison & Abramovitz was hired by Alcoa to invent new ways of using aluminum in building, developing architectural elements such as aluminum windows, doors, siding, and ductwork.[45] The most promising product that came into its own after World War II was the structural application of aluminum in office buildings—the curtain wall. Alcoa sought projects involving what it described as "monumental architecture"—large skyscrapers such as the UN Headquarters building that employed curtain wall systems or other types of aluminum skins. By 1955 Alcoa could boast that "more than 300 aluminum-skinned buildings were either finished or under construction."[46]

By 1952 the largest advertisement created for Alcoa was its thirty-story headquarters in Pittsburgh, designed by Harrison & Abramovitz. The skin consisted of an aluminum panel system, Alumilite, that did not require welding; panels were simply riveted into place. As Reinhold Martin noted in his book *The Organizational Complex: Architecture, Media, and Corporate Space*, many corporations in the postwar period began to use the exterior surfaces of their new modern headquarters to represent the company's product. The Inland Steel Headquarters fashioned a stainless steel curtain wall; Reynolds Metals Headquarters was shrouded in aluminum screens; Corning Glass sported a glass facade. The postwar era marked the appearance of a phenomenon in which the entire surface of corporate buildings corresponded with the advertisements published in *Fortune* and *Newsweek*. Martin illuminates the metaphorical connection between the metonymic surface of an office building and the surface of the magazine advertisement:

> In the work of SOM and their peers in the United States after the Second World War, the semantic message was as direct as the advertisements produced in the national and trade weeklies.... And as total construction volume increased fourfold in the United States between 1946 and 1956, for a brief period the rhetorical "industry" of modernity... coincided with the construction industry. Although this industry was not to remain its own client for very long, the convergence afforded architects like Bunshaft opportunities to exploit, with acute matter-of-factness, the temporary transparency of signifier to signified in the gleaming skins of the office building.[47]

Alcoa was very much a part of this moment in which the building's skin acted as both sign and signified, utilizing the surface of its headquarters as a giant billboard for its product. If, in early Alcoa advertisements, ALUMINUM IS WWII AIRPLANE, in postwar advertising, ALUMINUM IS BUILDING.

Aluminum is, at one point along its production cycle, a liquid, and therefore can be molded into a variety of shapes and objects.

The Alcoa Building, Pittsburgh, Penn., designed by Harrison & Abramovitz, 1952

While this "fluidity" is a physical attribute, this characteristic "need to be shaped" also lends itself to use as a metaphor. The phenomenon Martin described was part of the larger metonymic maelstrom that would provide the basis for Alcoa's marketing campaign, FORECAST.

Ketchum, MacLeod, and Grove, the company responsible for creating Alcoa advertisements, equated their FORECAST campaign with Harrison & Abramovitz's Alcoa Headquarters, in which advertisement and building were constructs of "ideas":

> [FORECAST's] principal objective is not to increase the amount of aluminum used today for specific applications,

> but to inspire and stimulate the mind of men, in much the same way they were inspired when the Alcoa Building was created. The Alcoa Building could have been built of brick and stone—it would have been a thoroughly efficient office building. But envisioned in this structure was something bigger and more important than an office building. The result stirred men's imaginations. FORECAST is built, not of brick and stone, but of ideas. It stimulates men to use aluminum in new ways—just as the Alcoa Building did.[48]

Architecture has always been identified with the innate ability to reify ideas or give structure to a series of concepts. Through the lens of the advertising world, Ketchum, MacLeod, and Grove understood the Alcoa Headquarters as being the ultimate signifier, where it not only represented the company and its product, but it also symbolized creativity and ideas. What the advertisers were attempting to illustrate by this passage was the ability of architects and designers to create metaphors using the ubiquitous medium of aluminum.

Ketchum, MacLeod, and Grove devised Alcoa's FORECAST program, a multi-faceted marketing campaign that included a series of advertisements in weekly magazines, a two-volume industry periodical titled *Design Forecast*, followed by the creation of a new type of showroom. The first phase of the campaign entailed the commissioning of twenty-two design projects by well-known and accomplished designers such as Charles Eames, Paul McCobb, Harley Earl, Eliot Noyes, Alexander Girard, Isamu Noguchi, Jean Desses, Marianne Strengell, Garrett Eckbo, and Herbert Matter. Each was charged with designing a product using Alcoa aluminum. The products created ranged from a ball gown made of aluminum fabric, to a garden patio, to a satellite-shaped stereo. The objective was not for Alcoa to eventually manufacture these items; as Alcoa president Frank Magee explained in the first issue of *Design Forecast* in 1959, "Our own industrial design group was established for the sole purpose of aiding industrial designers in *their* projects—not in any way to compete with the designer!"[49] By enlisting star architects and designers to create new aluminum products, Alcoa hoped to inspire its manufacturing clients to copy the designs, specify Alcoa aluminum, and mass-produce the product, thereby increasing demand. At the beginning of the advertisement campaign in 1955, Ketchum, MacLeod, and Grove described the FORECAST designs for the advertisements as "set[ting] off concentric waves of influence. In an evolutionary economy, they are shaping original thinking in new directions. To customers, suppliers, future employees and investors, they are showing the wonderful, imaginative spirit of Alcoa."[50] The company hired a group of celebrated photographers such as Richard Avedon, Irving Penn, Bert Stern, and Leslie Gill to photograph the product designs. Their photographs were transformed into advertisements and published in popular magazines such as *Time*, *Newsweek*, *The New Yorker*, the *Saturday Evening Post*, and *U.S. News and World Report*. The slogan that appeared on every advertisement read, "There's a world of aluminum in the wonderful world of tomorrow."[51] By employing the military tactic of forecasting and the rhetorical device of metaphor, designers and architects recruited by Alcoa orchestrated the creation of a future postwar domestic realm, with aluminum as their medium.

OVERKILL

The advertisements produced by Alcoa as part of its FORECAST campaign were based on practices at the RAND Corporation employed by Herman Kahn, a scientist, military strategist, and founder of the policy think tank the Hudson Institute. In 1948, while working at the RAND Corporation, Kahn first applied the term "scenario" in military planning for nuclear warfare. He defined "scenario" as "a hypothetical sequence of events constructed for the purpose of focusing attention on casual processes and decision points," [52] with some scenarios "emphasiz[ing] different aspects of 'future history.'"[53] In his book on the possibilities and outcome of war, *Thinking about the Unthinkable*, Kahn described the outcomes of first versus second nuclear strike scenarios, inventing the terms "overkill" and "megadeath."[54] A scenario can also be characterized as "an exploration of an alternative future," or as "an outline of one conceivable state of affairs, given certain assumptions."[55] The Alcoa FORECAST campaign functioned in a similar way, in which the goal of the designer was to create a series of scenarios depicting postwar domestic bliss projected into the future, using aluminum products.

One of the first advertisements in Alcoa's campaign publicized the creation of a ball gown in aluminum fabric. The photograph, taken by Avedon, featured the American model Suzy Parker emerging from a car in front of the "elegant Paris restaurant, Marigny," with the Eiffel Tower, the ultimate symbol of Paris and its cultural superiority, in the background. The caption read, "There's a world of aluminum in the wonderful world of tomorrow...where the coat that keeps your daughter warm...and the gown that turns all eyes to her...will be cut from aluminum cloth as silky as a butterfly's cocoon."[56] Published in *Newsweek*, *The New Yorker*, and *Fortune*, the ad was directed at the business executive of a manufacturing company who had the ability to specify aluminum fabric for use in one of his products as well as the cultural capital to buy the latest fashions for his "little girl." The photograph acts as a sort of cultural hijack, where the images featuring Parisian icons are intended to appropriate the cultural sophistication of French life onto a new, futuristic American company and its products.

Another early advertisement in the FORECAST campaign appealed perhaps to a more relaxed authority figure. The photo, taken by Penn, contains a colored aluminum table designed by Noguchi. Objects placed on the table suggest a relaxing, romantic evening at home with a glass of wine, some fruit, nuts, bread, and a rose. This type of arrangement portrays a different type of life than the one that existed before and during the war. The table and its objects symbolized a life that contains simple pleasures and the leisure time to enjoy them. The playful blues and purples of the table reinforce the relaxed outlook of the new postwar generation and its aspirations for a bright, colorful future.

Some of the scenarios depicted in the FORECAST advertisements feature architectural structures designed for suburban recreation. Eliot Noyes designed a modular aluminum shelter in

Advertisement for Alcoa aluminum ball gown, designed by Jean Desses, Paris and photographed by Richard Avedon, as it appeared in *The New Yorker*, October 13, 1956

which backyard activities such as playing, barbequing, and gardening could take place. He left enough space under the awning to house the new Harley Earl-designed car, with its Alcoa aluminum fenders and grille.

In addition to designing Alcoa's headquarters, Harrison & Abramovitz designed a summer house based on the geometry of triangles, reminiscent of Alcoa's logo. The house, built on a type of large turntable, rotates to follow the arc of the sun. A company advertisement promised,

> You will pass your leisure hours at beach or mountain with all the comforts of home...in a summer house made of aluminum...roofed, walled and screened with aluminum to turn the sun and welcome the breeze...and take any shape or color to suit the landscape that wears it like a crown jewel.[57]

The American postwar life included not only leisure time but often a vacation house to spend it in. If a second home was out of the question, aluminum, in an accordion-shaped camp trailer, could accommodate the desire for recreation through the exploration of the country and its (revitalized) national parks across the country.[58]

Charles Eames designed what was deemed by Alcoa as a "moving sculpture" or "solar toy"; he described his invention as "a device that will do nothing." It was, in fact, an innovative working model of how solar energy could be converted into motion.[59] Most importantly, the solar toy served as an opportunity for social commentary, as Eames suggested: "This could be a good starting point for somebody wanting to make a design: to think first about what he wanted to make people aware of, and then to move toward the most effective and pleasing way of bringing this about."[60]

The solar toy could not be easily exploited and modified for mass production, but Eames's design sensibility pervaded many of the projects featured in the Alcoa advertisements. An aluminum view box designed by John Matthias was a trimmed-down, generic version of the Eames House. With panels of primary colors and a rigid aluminum frame, the view box was designed to accommodate outdoor recreation for the nomadic postwar family. "As portable as the tents of the Arabs," it would provide a sheltered retreat "any place that beckons on mountainside, meadow or shore."[61]

The advertising developed by Alcoa as part of its FORECAST program was institutional advertising, designed to improve the company's overall image. In 1960 Alcoa continued the program by producing the publication *Design Forecast*, which had a different mission. Edited by the company's in-house chief industrial designer, Samuel L. Fahnestock, the purpose of *Design Forecast* was to inform the people "who are doing the choosing and specifying" of aluminum's capabilities. In the second issue, Fahnstock rhetorically inquired about the future of design and designing for the future:

> What kind of planning for future products? Can we plan for an unknown future? If so, how?...We start this attack on tomorrow by insuring that we are doing everything in the

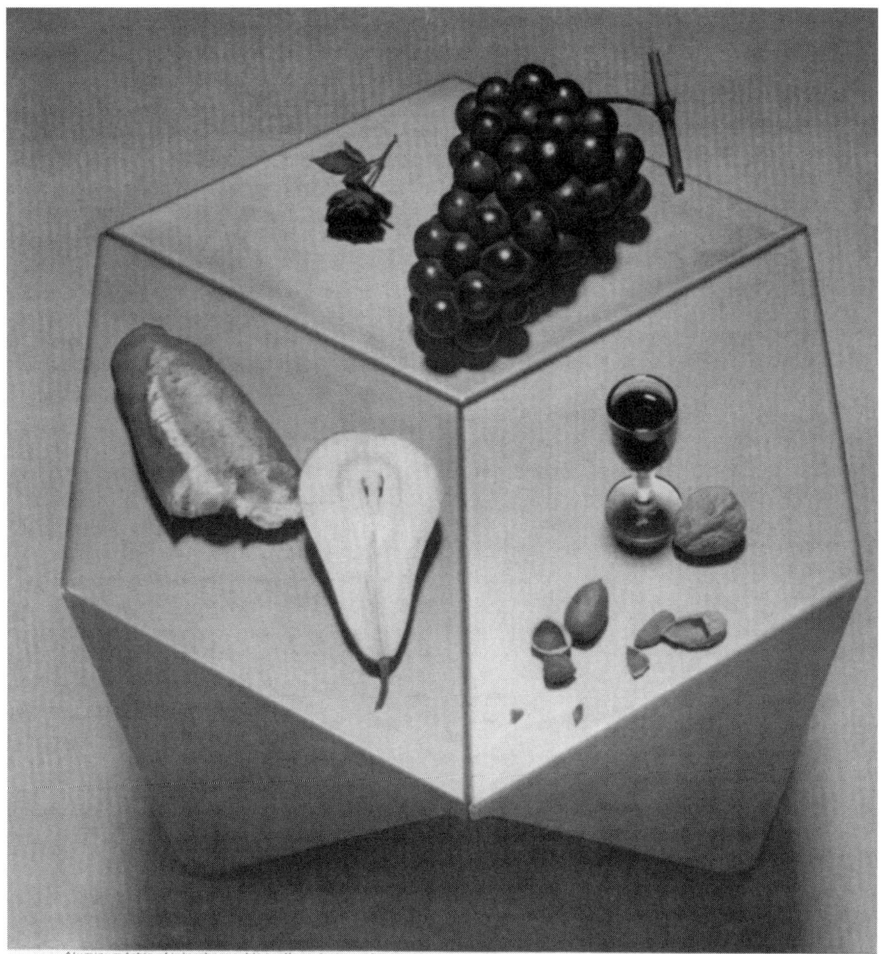

Aluminum table of interchangeable sections designed for the Alcoa collection by Isamu Noguchi. Photographed by Irving Penn.

FORECAST: THERE'S A WORLD OF ALUMINUM IN THE WONDERFUL WORLD OF TOMORROW
...where you will style rooms to your whim of the moment...using sectional aluminum furniture of myriad textures, colors, finishes and forms...in arrangements as endless as the patterns of a kaleidoscope.

Advertisement for Alcoa end table, designed by Isamu Noguchi and photographed by Irving Penn, as it appeared in *The New Yorker*, May 11, 1957

FORECAST:

THERE'S A WORLD OF ALUMINUM IN THE WONDERFUL WORLD OF TOMORROW ... WHERE YOU WILL FIND PROTECTION FROM SUN AND RAIN ... AND CANOPY YOUR CAR, PATIO, POOL OR PATCH OF BEACH ... UNDER WIDE-SPANNING STRUCTURES OF ALUMINUM AS AIRY AND COLORFUL AS THE SKY ... ALUMINUM STRONG ENOUGH TO WITHSTAND GALES AND YEARS ... ALUMINUM SO EASY TO SUPPORT, YOUR SHELTER SEEMS TO FLOAT ON FINGERS OF LIGHT.

ALCOA ALUMINUM Aluminum Company of America, Pittsburgh

Modular aluminum shelter designed for the Alcoa collection by Eliot Noyes. Photographed by William Bell.

Alcoa aluminum modular shelter, designed by Eliot Noyes and photographed by William Bell, as it appeared in *The New Yorker*, July 25, 1959

FORECAST:

There's a world of aluminum in the wonderful world of tomorrow

. . . where any place that beckons on mountainside, meadow or shore will become your sheltered retreat . . . because aluminum will travel with you . . . standard shapes of light, strong, weatherproof aluminum that become airy, expansible shelters as kind to the landscape as sunlight . . . as portable as the tents of the Arabs.

Aluminum Company of America, Pittsburgh

Alcoa aluminum view box, designed by John Matthias and photographed by Ted Castle, as it appeared in *The New Yorker*, October 24, 1959

The Forecast Accordium Camp Trailer, designed in aluminum for the Alcoa collection by Henry Glass. Photographed by J. Frederick Smith.

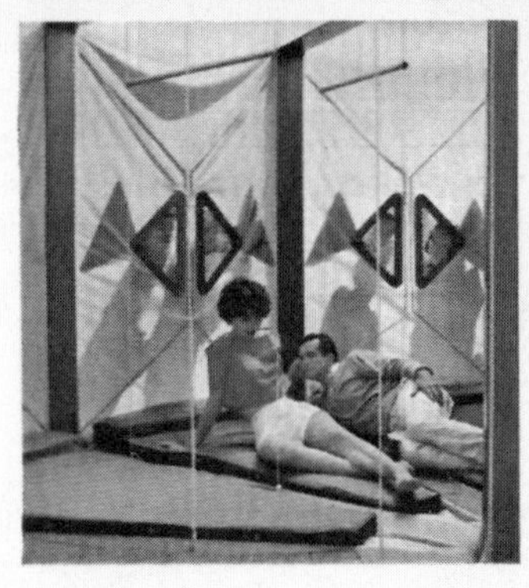

Forecast: There's a world of aluminum in the wonderful world of tomorrow... where shelters of aluminum will fold like an accordion to accompany you anywhere... trailers of light, strong aluminum sections that wed with other materials . . . carry accouterments of work or leisure... withstand road and storm ... and unfold where you will with peacock glory.

ALCOA ALUMINUM

FORECAST accordion camp trailer, designed by Henry Glass and photographed by J. Frederick Smith, as it appeared in *The New Yorker*, September 17, 1960

> best possible manner today....Our depth of focus has to be increased, so that the detail in the foreground becomes the point of reference for observing and reckoning with a clearly seen future that is literally rocketing onto our desks, drawing boards, and plant floors.[62]

Alcoa President Frank Magee stated in the introduction to the inaugural issue that the publication was directed toward an audience of designers, architects, and design-minded business executives. The objective of the magazine would be to ultimately "outline why Alcoa is vitally concerned with the design field."[63] He explained that, historically, aluminum had played the role of displacing other materials; objects that were once cast or forged in steel or iron, from teakettles to airplane propellers, were now made with aluminum. According to Magee, the magazine, like Alcoa Headquarters, would encourage the trend of aluminum replacing other metals by providing a "means of inspiration" for designers, to assist them in imagining new applications. Thus, while in the past aluminum simply displaced other materials, designers and architects—with the help of advertisers—would appropriate new meanings (or, at least, inspire other designers and architects to appropriate new meaning) for their ubiquitous and malleable product.

Originally conceived as a serial publication with subscribers, *Design Forecast* lasted for only two issues. The contents of these issues can be broken down into three categories, reflecting the technical, artistic, and professional concerns facing industrial designers.[64] In most of the articles, an architect or designer predicted a future trend. In "Design Trends: Aluminum in Furniture," for example, Paul McCobb and Charles Eames predicted that aluminum would replace wood in most furniture manufacturing, and *Harper's* editor John Fischer forecasted the return of the American home and its importance in postwar culture.[65]

Predicting the future, or forecasting, was a common thread throughout the entire publication. Vice president and general manager for the Major Appliance Division of General Electric Charles K. Rieger and company designer Arthur N. Bec Var presented their viewpoints on design in a twin set of articles titled "Appliances for Tomorrow." Rieger wrote,

> Anyone in the appliance business who dreams about the challenge years ahead must keep one point firmly in mind: the manufacturer who succeeds will be the one who makes things people want.... The industrial designer who elects to help an appliance manufacturer solve this problem must lead in the particular art of anticipating what people will want even before they know that they will want it.... He must exercise it in cooperation with scientists who can keep him abreast of materials and processes that are about to become available. He must exercise it in cooperation with economists who can predict income and spending trends, and with sociologists who can suggest the directions of family growth and activity, and population shifts.... Practiced in this way industrial design becomes a resource of management to insure the company's future growth.[66]

The industrial designer, like the military strategist, should consult with other experts in order to predict the future needs of American consumers. According to Rieger, the strategies of a well-informed industrial designer were analogous to the tactics of a RAND forecaster.

Walter Dorwin Teague answered the rhetorical question lodged in the title of his article "When Should the Industrial Designer Enter the Picture?" He began by enlarging the definition of the designer to include not just the design of product but everything that constituted the "corporate image." Teague asserted,

> Many design groups, both independent and corporate, render a variety of services ranging over product displays, exhibits, architecture, offices, sales outlets—everything that goes to make up the "corporate image," or create a distinctive corporate personality; everything, that is, but advertising, which is a specialized service itself.[67]

Tellingly, Teague concluded by alluding to the business plan as war preparation, "Let [the client] map a long-term campaign that will win not a mere battle but the war, and let him follow its strategy without deviation. Both he and his customers will profit, and public taste will benefit."[68]

One of the most intriguing articles published in *Design Forecast* was actually a transcript of a symposium held at an Alcoa corporate retreat, in a lodge in the Tennessee mountains.[69] The symposium, organized by Alcoa to discuss the future of design, mimicked the brain-storming Delphi technique practiced at the RAND Corporation. During the 1950s, RAND used the term Delphi to define brainstorming sessions conducted to "obtain the most reliable consensus of opinion of a group of experts... by a series of intensive questionnaires interspersed with controlled opinion feedback."[70] The characteristic that distinguished the Delphi technique at RAND from traditional corporate meeting techniques was that the participants were qualified experts and consultants, usually civilian scientists or researchers from the military, academia, or industry. The initial application of the Delphi technique at RAND was directed toward "the selection, from the point of view of a Soviet strategic planner, of an optimal U.S. industrial target system and... the estimation of the number of A-bombs required to reduce the munitions output by a prescribed amount."[71]

The symposium convened to discuss the theme "Industrial Design: 1950–1960–1970." The group of speakers, like the consultants at RAND's Delphi sessions, included a variety of experts from the field of industrial design, manufacturing, and marketing. The meeting was attended by designers and businessmen (and some who successfully filled both roles) as well as figures such as *Industrial Design* magazine's editor Ralph Caplan and publisher Charles E. Whitney. All of the participants were male, and all had attended college, some graduating before the war and some after. Unlike meetings held at RAND, no one from academia or from government was present, but the symposium did imitate the Delphi technique of gathering consultants from different areas of the design field.

The discussions were informally led by Caplan and Fahnestock, who attempted to base the conversation on four general questions:

"What is the creative man?"; "What are society's values in respect to things?"; "What is the relationship of design to 'physical research'?"; and "What must the designer be doing tomorrow?" The discussion did not address these topics equally, however; the majority of comments seemed to respond to the latter two questions regarding research and "design for tomorrow."[72]

Douglas Kelley, a designer, defined the goal of his profession as injecting a certain theory as well as function into a product: "The designer must contribute to management a philosophy of living that perhaps pure industrial society cannot comprehend: a manner of living, a unifying of products to provide a value of life, a combining of products and functions to give greater values in the marketplace and to the user."[73] What Kelley suggested for the designer was not unlike the role of RAND forecasters in creating military scenarios. In creating "a manner of living," the designer illustrated for the consumer a scenario of how to use the product, infusing meaning and value into the product by placing it within the domestic landscape of the postwar American utopia. Bill Snaith, president of the Raymond Loewy Corporation, attempted to create an economic scenario for the average consumer, asking industrial designers to strategize according to this prediction:

> Think of 1970, and what the economists say we will have. The average income will be $7500, and discretionary income will be increased at least fifty percent.... With a large discretionary income, people will either buy more of goods of average quality, or will insist on higher quality goods. What can the designer start to do now to aim his employer or clients toward this higher standard for tomorrow?[74]

Reflecting the design industry's approach as a whole, the rhetoric of this conversation ranged from military speak to organizational or business jargon. Snaith continued to describe the operation and structure of his design firm in terms applicable to the organization of RAND:

> Our firm is doing three types of forward programming of product and distribution research that I think indicate the

> changing and future role of the designer. First, with one organization we are working on a "program for invention," planning in terms of what could and should be developed for tomorrow's products. This is then analyzed in terms of things that are possible in the near future, and things that must wait for further developments and further contributions of research.[75]

The techniques applied to the development of jet aircraft design, missile deployment strategies, and thermonuclear war scenarios were also applied to the design of postwar domestic goods, such as washing machines, ovens, and kitchen blenders.[76] The designer Brook Stevens confirmed this idea when he concluded the gathering with a comment to Snaith: "Bill, you have become a situation planner."[77]

FORECAST JET: FLYING SON-OF-A-SIGNIFIER

Despite the effort invested into advertising and marketing campaigns in the 1950s and 1960s, the market would remain problematic for Alcoa. The biggest obstacle the company would encounter was a predicament that could not be alleviated by any designer or architect. In the fall of 1965, President Lyndon Johnson initiated a rollback on the price of aluminum as part of an overall attempt to mitigate inflation exacerbated by the Vietnam War. By then the stockpile bulged with a surplus of aluminum, thus giving the government great leverage with which to control aluminum prices. Alcoa's management had realized the possibility that one day they would have to buy back the government's surplus at government-mandated prices; that time arrived in 1965.[78]

Alcoa's aluminum production, and national acclaim, had reached its greatest heights during World War II and in the years that followed had steadily declined. The Vietnam War brought to a close the chaotic period of aluminum production, with the industry grasping at industrial design and advertising to increase demand and profit. The aluminum industry began its downward spiral in the late 1960s and early 1970s, when the postmodern rejection of monumental architecture and the financial constraints created by the

The Alcoa "flying showcase"—the corporate jet as showroom, 1965

oil crisis rendered aluminum both unfashionable and extremely expensive.

The final phase of Alcoa's FORECAST campaign involved the creation of the FORECAST corporate jet, a flying showroom that displayed the company's products and services. The aircraft, a DC-7CF purchased by Alcoa in the spring of 1965, was transformed into an "aeronautical ambassador of aluminum." Designed "to expand the markets for aluminum, increase awareness of Alcoa, and dramatize the company's creative achievements,"[79] the plane was flown by a three-man crew and hosted staff from management, sales, research, and development, depending on the specific objectives of the flight. While on the ground, the plane could expand to form an exterior reception area.[80] An Alcoa promotional brochure described the area as "a blue vinyl-carpeted reception court formed by a semi-circular screen of aluminum beads.... resembl[ing] chain-mail armor." The visitor would board the plane via a "grand staircase, a cantilevered 12-step spiral centered on an aluminum flagstaff. Each step is of aluminum honeycomb construction."[81] Once inside, the visitor entered a conference lounge where the interior was fashioned with as much aluminum as possible—from "rosewood inlaid with woven aluminum panels and aluminum strips finished in shades of bronze by Alcoa's Duranodic 300* process" to "wall, furniture and carpet fabrics contain[ing] aluminum yarns." A dominant feature of the lounge was a suspended light fixture consisting of a "deliberately ornate" grid-work pattern of aluminum sand-casting. Light was diffused through colored glass inserted into the grid. One of the art pieces decorating the lounge, constructed of 78 aluminum extrusions, was titled *Extrusion City* because of its "striking resemblance to a metropolitan skyline."[82]

The center section of the plane consisted of a 40-foot showroom area devoted to product display. Actual product samples, cross-sections of parts, models, photographs, and transparent color illustrations were mounted or hung from a modular aluminum display system. The fixtures were flexible enough to allow for quick changes, so that the display could be tailored to the clientele.[83]

On the display shelves, between the aluminum beer cans manufactured for Anheuser-Busch and the children's anodized aluminum lunch boxes sat... architecture. As one aluminum object among others,

Alcoa FORECAST jet, rendering of exterior showroom space, 1965

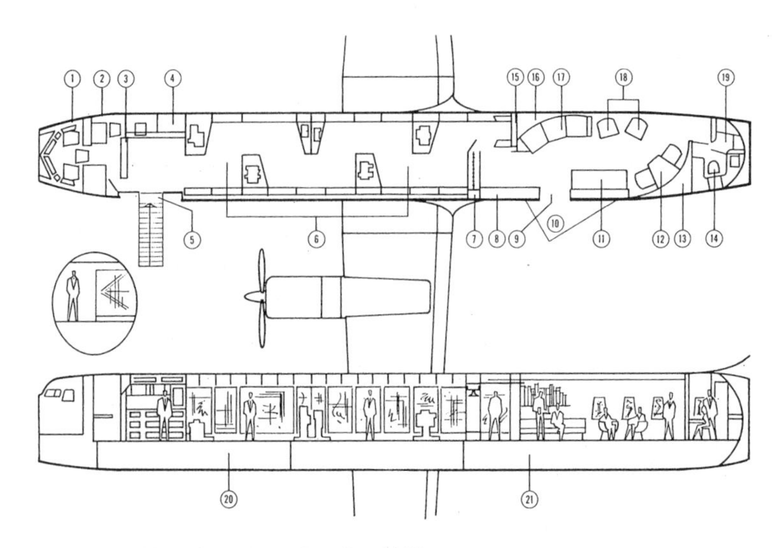

Alcoa FORECAST jet, floor plan and section, 1965

the curtain-walled architectural models stood, as if waiting to be chosen over the domestic goods produced from aluminum. Within this setting, aluminum's literal and metaphoric malleability reduced monumental architecture to a standard commodity.

Described by Alcoa President John Harper as representing "a new adventure in communications techniques,"[84] the entire airplane, from the aluminum stairways to the exterior skin, was a flying advertisement, a flying signifier, for Alcoa. Like the Alcoa headquarters building that preceded it, the FORECAST jet functioned both as a sign and signifier of its product as well as a metaphor for the new postwar corporation. The material had once again metamorphosed into another symbol; the product previously promoted as ALUMINUM AS BUILDING in the early postwar period had transformed into ALUMINUM AS CORPORATE JET.

The "aeronautical ambassador of aluminum" was the culmination of a Cold War semiotic tempest that began during World War II and continued in the years to follow through the work of the military-industrial complex. The corporate jet—the ultimate symbol of corporate achievement—referenced the heyday of Alcoa's production of

Alcoa FORECAST jet, interior meeting room with aluminum ornament and fixtures, 1965

military aircraft during the war. Here the aluminum airplane—*the* war machine of World War II—transformed into a sleek communication machine in the Cold War marketplace. All of the displayed aluminum goods were contained within another type of metaphor. Just as fleets of World War II airplanes carpet-bombed cities in their effort to defeat the Axis forces, the FORECAST corporate jet would travel across the country, jettisoning its generic wares of aluminum cans and curtain-walled buildings onto American cities.

The FORECAST advertising campaign revealed the new role of designers and architects and demonstrated how their work could be reduced to creating an object specifically for the sake of creating an image. Like the ad man, the level of skill stopped at the level of surface—the surface of the magazine page.

Like Alcoa, the RAND Corporation would expand and continue to flourish. The "experts" at RAND led the country into the ultimate exercise in operations research and systems analysis—

Alcoa exhibition, mounted inside the FORECAST jet, illustrating the many skyscrapers clad in its aluminum

the Vietnam War—casting a dark shadow across the "wonderful world of tomorrow" forecasted by Alcoa.[85] RAND and other military think tanks such as the Hudson Institute are still in existence, continuing to mill out forecasts and scenarios for the next war. Perhaps the greatest danger the military-industrial complex poses is that over time, through the "domestication" of hothouse tactics, its practices and logic become invisible, controlling the most banal as well as the most important details of our quotidian lives. In our contemporary world, the logic of the military-industrial complex has become so pervasive and transparent that we can rarely identify where its realm of control begins and ends.

PLASTICS

STEPHEN PHILLIPS

On October 28, 1952, architects from around the nation gathered for a two-day conference in Washington to "dream boldly about the future."[1] As reported by Betty Pepis, participants declared that there would be "a Golden Age of American architecture" when and if architects learn "to take full advantage of new plastic materials" advanced during World War II.[2] Architects purported that through the use of new lightweight plastics, "upkeep and maintenance" and "dark corners" could be eliminated in housing. One architect presented the ideal vision of a house that could completely fold up and out of the way through the use of mobile plastic partitions. Another even suggested the construction of "plastic cities" built over tropical waters through the advent of high-strength waterproof plastics.

Monsanto House of the Future (MHOF), 1957

Robert K. Mueller, vice president of the Plastics Division of Monsanto Chemical Company, concluded the conference by claiming, "the future of plastics in building is limited only by our imaginations and the public acceptance of new concepts in living."[3]

During the 1950s, designer imaginations were brewing with possibilities for postwar construction in plastics; however, as Monsanto realized, if there was ever going to be any future to these fantasies, it was not going to happen without public interest and approval. To achieve this goal, on October 5, 1955, Monsanto executives authorized the proposal for their House of the Future project, as initiated by manager Ralph F. Hansen of Monsanto Plastics' Marketing Division. The Monsanto House of the Future (referred to by its creators as MHOF) was designed to enhance and guarantee the long-term viability of the company's new wartime plastics industry. To ensure peacetime applications of their products, Monsanto promoted both the need and the desire for a new paradigm of modern architecture practice—"plasticity."

WARTIME PRODUCTION

As reported in 1951 by Frank Curtis, director of Monsanto's long-range development program, the smooth transition made between peacetime and wartime applications of the company's chemical manufacturing lines was considered a model of success. Prior to World War II, the chemical industry rapidly expanded production of basic manufactures that could be easily converted to wartime uses "merely by changing the destination of a tank car."[4] Both the chemical industry and the Chemical Warfare Service were aware that certain kinds of ammunition and high explosives would be in great demand during a future war, and companies such as Monsanto focused the peacetime efforts of their chemical divisions on manufacturing domestic products that had base materials similar to those of explosives.[5] With domestic sales of their antiseptics, detergents, and cleaners setting all-time records in 1939 and 1940, Monsanto was able to swiftly redirect its expanded peacetime production of the base materials phosphorus, sulfuric acid, chlorine, trisodium phosphate, caustic soda, and phenol to the production of various explosive materials in order to meet peak World War II demand.[6]

Unlike the chemical industry, the plastics industry was not prepared for an easy transition from domestic to wartime production. A company report on Monsanto's World War II efforts acknowledged that "while conversion to war in the old organic-chemical line was by and large not too difficult, the Plastics Division had an extremely difficult time."[7] Plastics had not been developed during peacetime with wartime applications in mind. Frustrating Monsanto's efforts to support the war was the fact that the defense industry was unfamiliar with the potential uses of synthetic chemicals developed and used domestically. When World War II broke out, Monsanto's major products consisted of safety-glass plastic, Resinox molding compounds (used to make plywood), Cellulose nitrate sheets and rods, phenolic resins (used in jukeboxes), and polystyrene (formed into novelties and gadgets).[8] As remarked by Curtis, "Practically none of [these products'] uses were approved for wartime."[9]

With traditional materials such as metals predicted to fall quickly into short supply during the war, the demand for plastics could prove dramatic. R. D. Dunlop, operating chemist of Monsanto, observed, "the first reaction after Pearl Harbor was that plastics would jump into their own."[10] A problem arose, however, with the replacement of traditional materials with plastic. Dunlop noted,

> When we saw the requirements that [plastics] must fill, it also became apparent that the job was not easy and a lot of hard work was required. The aircraft industry very quickly and emphatically told the plastics industry that plastics in aircraft for structural purposes were a possibility but that no one knew enough about them to safely design an airplane incorporating them. The electronic industry said that some of our materials, particularly polystyrene, had very admirable properties, in some respects, but they were not quite sufficient to fill every requirement. . . . No one in the plastics industry had worried a great deal about specifications.[11]

Glass fiber reinforced plastic radome (GRP), installation

Unfortunately for the plastic industries, there had been little effort during peacetime to develop materials and methods that could meet the high standards of wartime production. Plastics developed by Monsanto had not been designed for their structural integrity: polystyrene, which had excellent electrical properties that could be utilized with radar, softened at a low temperature; nitrate, which could be used for aircraft glazing, burned too readily and was easily affected by the heat of the sun; Resinox, designed as a general purpose wood resin, was suitable for low-impact applications only; and one of Monsanto's most significant peacetime products, Buvtar, used in safety glass plastics for automobiles, was considered relatively useless for wartime production.

Between 1941 and 1943, Monsanto worked quickly to develop plastics better suited to war. They did considerable work to determine the "physical-mechanical" properties of plastics so that they could provide data to the Armed Forces and various other manufactures of war materials.[12] They worked to create more viable plastic products and, as needed, built extensive new production facilities.[13] Of great success was Monsanto's development—in cooperation with Owens Corning Fiberglas—of Thalid resin for bonding glass cloth. The brand name product, Fiberglas, was used extensively for light, non-shattering, flexible Doron plastic armor suits, and Thalid resin and glass cloth were used in structural aircraft parts such as radomes.[14]

During the war, new plastics significantly advanced developments in radar technology, and radomes made almost exclusive use of the new composite Glass Fiber Reinforced Plastics (GRP). Proving to be transparent to radio waves as well as lightweight, strong, and weather resistant, GRP allowed radar on planes and ships to be protected, operational, and mobile. Lockheed Corporation manufactured radomes with the newest technology, using an interior and exterior reinforced structural tension shell of GRP held together with interior structural foam plastic—Lockform. "Foamed" into place between the inner and outer skin, Lockform's key ingredient was plastic. Lockheed used this technique of foamed-in-place, double-shell construction for other structural aircraft parts, such as the ailerons and rocket exit doors of the Lockheed F94C *Starfire*. *Monsanto Magazine* editors noted in the article "The Nose that Sees" that this new manufacturing technique made possible "a great saving in man-hours" as "an aileron, which formerly had many ribs and doublers inside and hundreds of rivets to hold it together, now has almost no ribs and few rivets."[15] Through the use of composite plastics of tension strength construction, the industry saved valuable time and materials in the manufacture of structural components for aircraft.

The technology devised to sandwich an interior core between structural, high-strength tensile surfaces was first developed in plywood. Using a plastic resin-coated balsa wood or foam core, the British manufactured the *Albatross* and the famous RAF *Mosquito* in 1940.[16] Monsanto resins added strength and durability to this original technology. As these resins were boil-proof, corrosion-proof, and waterproof, they could increase the time the plywood could be soaked in salt or fresh water, thus adding durability to the finished product. Setting at a normal 70-degree room temperature, these new high-performance resins were easily accommodated in the manufacturing process. The application of these resins was extended to the construction of wooden boats, including the PT–boat type, as well as plywood fuel tanks.[17] Able to be mass-produced in seamless, thin, lightweight, continuously curved units that would not "create wind resistance or invite leaks," plastics provided the aircraft and shipping industries with a valuable new material from which to produce strong, durable, aerodynamic, waterproof, formed structures.[18]

The use of composite glass fiber reinforced plastics in the aircraft and shipping industries only further enhanced the benefits seen with plywood. In 1944 the U.S. Army built and tested the first airplane fuselage made of glass fiber cloth with a balsa wood core laminated with Monsanto's X–500 resin.[19] Test flights pronounced that the fuselage made of GRP achieved tensile strengths up to 46,000 pounds per square inch and were effectively stronger, and certainly lighter, than standard metal sections of aluminum or steel.[20] Plastic resins combined with glass or wood reinforcement showed great potential for long-term structural applications within the aircraft and shipping industries, and Monsanto was particularly interested in developing these technologies for future production.

Molded plastic and plywood products appeared to have distinct advantages over metals used previously for manufacturing large industrial products. U.S. molding methods using small press-formed or cast sheets of aluminum steel required that the materials be cut, fitted, and then fixed together to achieve complex shapes—a laborious and inefficient process that produced a great deal of scrap. These methods were inadequate to meet the increased demands of war. Large, structural plastic- and plywood-shaped sections provided greater continuity with fewer connections between parts, which was important for time and material efficiency. Monsanto resins contributed significantly to the effort to create new methods to achieve large-shaped products with greater structural integrity.

The plastics industry took great pride in their wartime accomplishments. Plastic protected the Allies against moisture, grease, dirt, saltwater corrosion, bullets, noxious gas, and even atomic fusion.[21] Companies such as Monsanto had created lightweight, mobile, time and material-saving, mass-producible plastics of continuous construction. They had proven themselves capable of developing a variety of new materials and technologies that were strong and durable enough to withstand the test of war.

When World War II came to an end, industry needed to shift production back toward domestic applications, and Monsanto would have to face another challenge: how to adapt these new plastic products, specifically designed for war, to peacetime production.

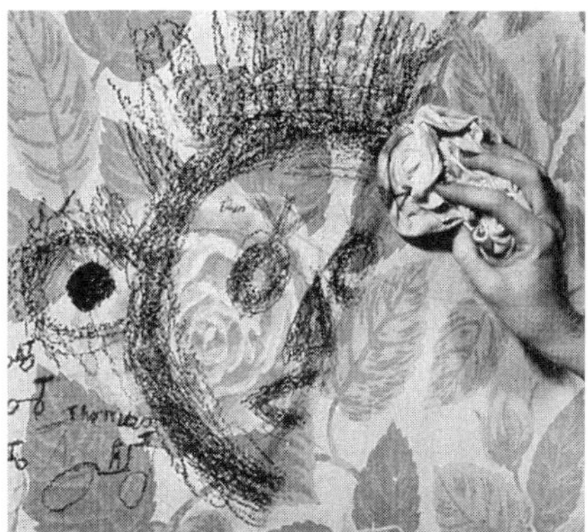

Illustrations from the January 1946 *Life* magazine article "Indestructible Room: New Plastics Protect Walls, Furniture and Rugs from Ravages of Kids and Dogs"

DOMESTIC PLASTICS

In 1943, *Newsweek* announced in their article "Test-Tube Marvels of Wartime Promise a New Era in Plastics" that industry would create a "plastic postwar world."[22] The wartime accomplishments of the plastic industry were presented to the public in popular magazines as the great hope to ensure the financial future of an expanding, post-World War II economy. There was a general consensus that no other innovation offered "such promise for rebuilding our war-torn industrial economy."[23] Just as plastics had proven to be instrumental in the war effort, so too would they be at home, after the war.

Some in the plastics industry were very concerned that the historic record of inferior plastics used before and during the war had tarnished the material's reputation. But a clear sense that plastic could benefit and even transform life at home emerged before the war ended. *Newsweek* speculated,

> For the postwar world, there are promises of plastic houses, of plastic private airplanes, of thousands of other articles that will heighten the comfort of everyday living. Plastics may be the key to a new industrial era. . . . They promise the production of basic materials tailor-made to fit the finished products.[24]

With plastics, a new world could be molded into the form of our desire. As the war ended, industry quickly shifted its attention to bring about this new synthetic world.

The media made clear how plastics could transform everyday domestic life: it could protect America at home on the domestic front using the same methods, materials, and techniques developed for war. Articles in *Life* magazine and *Better Homes and Gardens* announced that it was domestic architecture that now needed to be protected.[25] *House Beautiful* devoted its entire October 1947 issue to plastics, promoting its benefits to ensure a safe household. Titled "Plastics: A Way to a Better More Carefree Life," the issue featured an image of plastic wallpaper being drawn on by children and easily wiped off by Mom.[26] This was a new, postwar Mom, at home, surrounded by her growing family. Newly developed plastics promised her a chance to "forget [she had] children" and live an "elegant, carefree life."[27]

Initially, plastics found their greatest use in the bathroom, laundry, and kitchen, "where the going is rough."[28] Easily cleaned, products like Formica and melamine cabinetry, vinyl flooring, polyethylene bottles and bowls, and phenol molded accessories proliferated in postwar years. "Damp-cloth" consumerism seemed the perfect weapon to fight the known enemies of dirt, grease, and grime in the home; plastic manufacturers and designers understood what products were needed for that war and were prepared to fight it, with little need to reinvent basic plastic manufactures or their fabrication processes. Durable, waterproof plastics were readily advertised to the public alongside the chemical industry's cleaners, detergents, and antiseptics.

Housework would become so easy that there would now be more luxury time to shop for glamorous synthetic outfits and cosmetics. The Monsanto advertisement "From finger-tips to wing tips" maintained that just as man had used plastics during World War II in the "conquest of the air," so woman could now use plastics at home in the "conquest of man!" Plastic products developed for the wartime aerospace industry found application in women's clothing and cosmetics.

"From finger-tips to wing tips," advertisement by Monsanto Chemical Company, 1946

Plastics had become the newest fetish in the war of the sexes, and Monsanto was not shy to promote its contribution.

Husbands, whose sex appeal supposedly derived from being rough, tough, and dirty, were frequently portrayed in advertisements spending time outdoors, in the yard or in the garage. Fiberglass plastics accompanied them, particularly in California, where backyard "do-it-yourself" (DIY) enthusiasts put together everything from furniture, surfboards, and swimming pools to boats and small aircraft made from this versatile new material. GRP, which used simple lay-up techniques, required only glass fabric cloth, a mold, resins, and an idea. Once assembled, it could cure outside in the sun. Pre-manufactured, lightweight, inexpensive, transportable, structurally sound, and easy-to-assemble kits containing fiberglass plastic parts were made readily available to the suburban DIYer.

After the first fiberglass plastic car was built in 1946, men could presumably assemble their own motor vehicle in the backyard.[29] By the 1950s building your own car had become a *Popular Mechanic* phenomenon, as small shops developed prototype kits for the "Brooks Boxer," the "Scorpion," and the "Wildfire."[30] Each car held the promise of beating, stinging, or burning the driver's opponent—effectively, his neighbor—off the road. As Thomas Hines demonstrated, stylistically, the 1950s cars all had mean grills and headlights to aggressively defend their occupants on the competitive suburban streets.[31] Even Detroit gave into the fad, churning out a fiberglass plastic Chevrolet Corvette in 1953 using Monsanto resins, which lent mass consumer appeal to this DIY trend.

Artists and architects also recognized the potential uses for mass-produced fiberglass plastics. Charles and Ray Eames and Eero Saarinen were some of the first to experiment with these materials immediately after the war, inspired by their earlier work with plywood.[32] The Eameses utilized fiberglass plastic panels made from wartime surplus to form chairs.[33] They produced various designs in plastic using Monsanto resins. The article "Furniture for Moderns," featured in *Monsanto Magazine*, explored their simple chair designs for Herman Miller, manufactured at Glassform, Inc. in Los Angeles.[34]

Using glass rope fibers, they formed a fiber reinforced tensile mat filled with polyester resin that, when set between a matched-metal die, molded to the shape of a seated body. Lightweight and integrally colored, these chairs were easily stamped out, trimmed, and mounted onto metal legs for mass production.

Saarinen's Tulip Chair and the Eameses's La Chaise were perhaps the most structurally innovative fiberglass plastic chairs. The Tulip Chair appeared as if it were entirely made of plastic—supported on a central leg seemingly continuous with the chair seat and back. La Chaise used double tensile shell construction to achieve its greater strength and lightness, its exterior layers of fiberglass plastics sandwiching a resilient core of foam rubber and Styrofoam. The ethereal form was suspended effortlessly on spindly metal legs above the ground. Modulated to conform to the body in motion, the chair's amorphous plastic form provided multiple seating arrangements. La Chaise was designed to float like a cloud in defiance of the forces of gravity. It suggested the new fascinations of a mobile and temporal modern society.

A limit to the development of these chair designs—as well as to the development of cars, boats, and aircraft parts—was always the structural integrity of plastics. In the late 1940s and well into the 1950s, plastics failed to serve structurally and would seemingly never live up to the basic technical requisites of life safety. Fiberglass plastics offered much hope as a structural material, but it would take some time to meet government agency standards. Meanwhile, Saarinen's Tulip Chair relied upon a stand cast in aluminum that was subsequently sheathed in plastic to support the weight of a seated person, and the car industry maintained a steel chassis in their cars and returned to using steel bodies instead of plastic. Monsanto was concerned that "plasticity" would not continue to receive the public and financial support it needed to achieve its full potential. Therefore, the company took the initiative to make a substantial investment to secure the long-term promise of plastic and moved forward in a concerted effort to gain public and governmental support for their products.

THE MONSANTO HOUSE OF THE FUTURE

Although Monsanto had generally succeeded in transitioning wartime plastics to domestic production, by the mid-1950s it believed it was important to evaluate the extent to which plastics had made a marked impact on domestic life and to reassess what opportunities remained for future development. After the war, Monsanto had successfully developed Chemistrand Corporation to produce acrylic clothing and nylon tire cables. They manufactured All Laundry Detergent and All Dishwashing Detergent. They began to develop interests in food manufacture and the production of Krilium soil conditioner.[35] They also maintained interests in the aircraft and shipping industries and produced antiseptics, laxatives, and aspirin. Many of their small-scale domestic products were developed from the raw materials used in wartime industry, which could be easily diverted back to wartime production if the need arose. Monsanto had not, however, had much success developing large-scale plastics for domestic production.

Through research conducted under Director Ralph Hansen of the Market Development Department in the Plastics Division, Monsanto identified specific areas within the domestic economy that held promise for plastics. Hansen steered Monsanto's attention to the potential market in existing homes over twenty years old for the "do-it-yourself" and "please-do-it-for-me" market, which accounted for over $7 billion in home improvements.[36] Research indicated that consumers were most "susceptible" to being sold home improvement products when they first move into a home. With over 150 million people moving from one house to another between 1948 and 1953, Monsanto executives were interested in determining how to get them to spend their money for products made of plastic.[37] Plastic wallpapers, vinyl tiles, paints, and melamine furniture kits all proved well-suited to the home improvement market. A product like the "do-it-yourself" fiberglass plastic pool was the perfect example of a large-scale home building kit designed to be easily installed in the yards of existing homes.

Also of great potential was the new housing market. This market had been steadily declining in the 1950s due to "the small crop of persons of marriageable ages" born during the Great Depression, and this trend was estimated to continue through 1960.[38] By the 1960s, however, new housing starts were estimated to climb. "It is to this vast market," Monsanto proposed, "we should set our sights of advanced design, allowing the intervening period for the transition."[39] The company believed it necessary to develop "new ideas which will motivate the consumer desire into demand and finally into purchase." To that end, it identified that "the responsibility of the architects and designers [was] to create ideas based on the psychological desires of the consumer." As "consumers do not always know what they want and why they act," the marketing department proposed the use of "intelligent experimental design, such as prototypes" to garner their approval.[40] Prototypes—used for furniture, toys, and auto interiors—had proven a successful marketing tool in obtaining quick acceptance of new products by both consumers and manufacturers. Monsanto believed it was possible to create similar desire on a much larger scale through the development of a housing prototype. Company executives at Monsanto had observed a trend in the construction industry toward modern design in schools and office buildings particularly on the West Coast. They believed that in time, "the ultra modern home" in plastic would be accepted. The company directed its efforts toward influencing that potential market.[41]

In May 1954 Monsanto approached Pietro Belluschi, Dean of the Department of Architecture at the Massachusetts Institute of Technology (MIT), to assist in their marketing effort. MIT was selected due to its close "liaison" with the plastics industry.[42] Albert Dietz, a specialist in housing construction at MIT, served as chair of the Society of the Plastics Industry (SPI) Committee on Plastics Education and had experience developing plastic armor suits during World War II.[43] The Market Development Department in the Plastics Division of Monsanto provided a grant-in-aid to the Department of Architecture at MIT to organize a committee headed by Richard Hamilton, also at MIT. The committee worked with Dietz to produce and publish a document in 1955 that outlined current applications of plastics in the housing industry.[44] The document listed every use of plastics—from wet applied vinyl and acrylic roof coating and polyester film vapor barriers to the more obvious cabinetry laminates and toilet seats—and showed that plastics were in the household serving insulative, moisture protecting, and sanitary needs. It concluded, however, that they were not yet being utilized for structure.

Plastics had not yet made any major contribution to the building industry as structural materials. Fiberglass plastic and plywood sandwich panel construction, invented during World War II for the aircraft industry, had achieved only minimal success in transitioning to the housing industry. In addition to design and fabrication problems, projects proved difficult to finance and distribute, which limited their success in the marketplace. Other house typologies invented in the postwar years—such as Buckminster Fuller's "all plastic dome" and General Electric's plastic "dream house," as well as more conventional prefabricated systems—likewise failed to have significant impact.[45] Traditional wood and curtain wall construction dominated the industry since 1950, as noted by MIT in their 1958 summary report *Building with Plastic Structural Sandwich Panels*.[46] Plastics had made few gains as a structural component within any industry, including aerospace, as everyone had been "waiting for the role of the structural sandwich in the building industry to become better established before they made a major commitment."[47] Monsanto realized it was important to make a deliberate effort to cultivate the use of plastics as a structural element and invested in two extensive efforts toward this goal.

To develop and support interest in existing plastic material applications, on June 3, 1957, Monsanto opened its "brand-new" Inorganic Research Building at Creve Coeur Suburban Campus in Saint Louis, Missouri. The building was designed not only to house investigations in promising new synthetic materials but also to demonstrate the use of extremely conventional technological applications of plastics to the building industry. The architectural firm of Holabird, Root & Burgee designed Monsanto's new office building utilizing over eighty different applications of commercially available standard plastics.[48] In addition to demonstrating a wide variety of standard applications, the company's goal was to demonstrate the potential use of plastics in curtain wall construction. The architects designed and installed sandwich type curtain wall panels that used foamed-styrene cores with colored facing sheets of reinforced polyester resin. Monsanto wanted to propose how plastics could be formed into structural members for standard curtain wall construction that might eventually be able to support primary building loads.[49] However, the company recognized it might be difficult to manufacture plastics to be more structurally sound than steel, and they accepted that "hybrid" plastic skyscrapers, with steel skeletons and plastic walls and floors "so light that the framework can be thinner," might be more feasible within ten or fifteen years.[50] To promote this potential, Monsanto installed at Creve Coeur full louver-type acrylic windows, plastic-faced concrete blocks, and styrene extrusions fitted in steel-frame channels to support interior reinforced polyester partitions.

If Monsanto's Inorganic Research Building at the Creve Coeur campus exhibited conservative applications for plastics available in everyday commercial building, its proposal for the Monsanto House of the Future sought to push the possibilities of using plastics in hope of establishing a completely new architectural typology. Mueller remarked in a talk before the Building Research Institute in Washington, D.C., in 1954 that plastic held the potential to completely revise the "architectural index" of the time. He predicted, "plastics play a significant role in a new American style of building architecture because of inherent features of plastic materials and their adaptability in any type of design."[51] Plastics would be the material of the future and had the potential to redefine architecture and the habits of modern living. Douglas Haskell, architect and editorial chairman of *Architectural Forum*, argued in his September 1954 article that plastics would generate a "second 'modern' order . . . to which today's 'modern' will be just an antecedent."[52] This "second 'modern' order" was to be derived on the basis of plastics' "inherent features" and would signal a departure from the current trend favoring the manufacture of steel frame, mechanically fastened panel construction, which used relatively little plastic. Haskell remarked:

> Today's typical "order," as Mies van der Rohe says, is the skeleton frame.... Tomorrow's structure may be typically all "skin." Its skin may be formed to become its shell *and* its interior columns of cellular structure.... A single continuous envelope of a thin sandwich material may yield structure and enclosure; resistance to destructive forces from outside; solidity or porosity; control of light and view; insulation for heat and sound, color and finish—all characteristics we now impose separately.... Future buildings may be as thin as eggshells.[53]

Continuous tension–skin eggshell construction as presented in 1950s architectural discourse sought to promote an alternative building typology in contradistinction to the traditionally accepted modern practice of "skin and bone" architecture. This was not the first time continuous tension–skin eggshell construction was proposed in architectural practice: Frederick Kiesler had already conceived this ideal method of construction in his 1933 Space House project.[54] Kiesler came out strongly against rectilinear panel and frame architectural practices, but he never had technological or industrial support for his vision. After World War II, however, the plastics industry had clearly realized the potential to develop and exploit the technology behind this architectural concept, which well represented the material characteristics of plastic over those of steel.

Knowing Kiesler's work, Haskell promoted the "inherent" material characteristics of plastics that, he wanted to believe, were only further enhanced by claims made by the atomic industry. In a special report by *Architectural Forum* on "Building in the Atomic Age," information gathered from several experts in the field of atomic energy supported claims that plastics had "proven" likely to be stronger than steel and completely fireproof when exposed to nuclear radiation.[55] In the report, Dr. Phillip N. Powers, director of Monsanto's Atomic Project, supported the long-term manufacturing benefits of the use of radiation or fission products from new nuclear energy reactors planned by Monsanto.[56] The company had been a significant contributor to the Manhattan Project during the war, and as it continued to research and develop atomic energy, it promoted the potential use of atomic radiation to enhance plastics.[57] Enamored by Monsanto's claims, Haskell predicted:

> Chemical, electronic, [and] radionic manipulation [would become] the *dominant process* in "building," which [had] hitherto been dominated first by handicraft and later by mechanical joinery. In over-all shape, buildings created by this new extension of *monocoque* principles, already familiar in the construction of airplanes and storage tanks, may well harmonize still better with the world of ships, planes and hangers than with today's typical rectangular 'frame' buildings.[58]

Haskell hypothesized that in the atomic age a new modern architecture would utilize gamma fission radiation to develop the material characteristics of plastics for use in continuous, *monocoque* structural skins.[59] These skins would bring about the replacement of those mechanisms that had previously supplanted handcraftsmanship. They would prove more appropriate to the science and speed of the modern, atomic age.

To supplant the technological supremacy of the steel industry, Monsanto recognized that it needed to make significant advances in engineering and construction practices favoring plastics. In order to achieve this goal, Monsanto sought to derive an authentic technology particular to the "inherent properties" of plastics. Hansen, citing from the program brochure of the MIT 1955 Summer Conference on Plastic Housing, maintained:

> Because many of the inherent properties of plastics differ widely from those of the traditional materials of the building industry, designers, fabricators, and producers entering this field face many new problems.... As a result, it becomes important for [them] to develop a thorough understanding of the material's basic properties and of the potentialities of new engineering combinations.[60]

Monsanto developed its interests in the House of the Future specifically to resolve the problems of design and fabrication thought to be particular to plastic materials. R. C. Evans, Monsanto's Plastics

Division marketing director, explained, "the design [of MHOF was] intended primarily to prove-out architectural and engineering concepts utilizing the inherent properties of plastics and thus stimulate the use of plastics in achieving more satisfying ways of living five or ten years [in the future]."[61] Monsanto had hoped to instigate interest, gain support, and prove the potential of a new architectural vocabulary that utilized and justified continued development, production, and marketability of its plastic materials. In particular, it intended to promote the use of "plastic sandwich panels fabricated entirely of one material" as "the answer to the curtain wall problem."[62] Monsanto hoped to suggest, with its prototypical design developed through strategic marketing and well-targeted investment, nothing less than a new architectural typology.

Accepting the challenge to develop and explore the potential of plastic architecture, MIT architects and engineers sought to formulate new designs, methods, and technologies that "dictated a sharp break with traditional architecture."[63] Generous financial support for their project initially came specifically from Monsanto and from the Corning Fiberglas Corporation. Not surprisingly, MIT would eventually conclude that the parameters defined as "inherent" to plasticity that could provide a "sound 'envelope for living' in an infinite variety of contemporary forms" would be achieved with high-strength tensile skin technology using Fiberglas structural sandwich panel plastic construction.[64]

As Hamilton and Dietz began to investigate the design for MHOF, in conjunction with Marvin Goody, an assistant professor of architecture at MIT, they produced a formal statement on the project's architectural evolution. They effectively declared their hope to invent a plastic "aesthetic" that did not "degenerate...into the realm of substitute materials" and instead "design...their product according to the dictates of the material."[65] The architects were aware that they were dealing with a material, the shape of which could be anything from flat to completely amorphous, depending on the molding process involved."[66] As Hamilton noted, "this was clearly an instance in which the designers were faced with a freedom that was all too complete."[67]

Although plastics theoretically supported the "freedom" to be formed without limit—tailored to one's desire—the MIT architects believed designing "form for form's sake," by "determining the shape and stuffing the function of living into it was not going to solve the problem of the ideal home"; this process, they believed, could not "adequately solve the needs of a mass client and in all likelihood would resolve in manufacturing difficulties."[68] Instead, they insisted that "the ultimate form had to be one that was peculiar to the plastics fabrication process" and would be derived in response to its material effects.[69]

Relying on technological materialist claims, they promoted an architectural process that was governed by the appropriate use of materials over any aesthetic, programmatic, or functional requirements. They argued their point by asserting that plastics performed best in compound curve or shell structures because they could be "easily molded into thin hull-like components" using the minimum amount of materials.[70] Curved, statically indeterminate forms achieve rigidity and stability from their shape and thereby theoretically use less material by weight to achieve the same stability as flat, rectilinear structures. This works for all materials, but plastic's malleability can achieve compound curves more readily. As plastic is typically lighter than most construction materials, there are also structural benefits to be gained by its use in curved forms.

It became clear, however, that when used in more traditional, rectilinear, statically determinate forms, plastics had a much more difficult time than other materials in achieving the rigidity and stiffness necessary for structural purposes. As Dietz realized, plastics' ability to achieve the rigidity and stiffness needed to withstand deflection loads while still maintaining (like steel) "plastic flow or yielding" in areas of concentrated stress would determine its applicability in the 1950s.[71] As plastics were developed for rigidity, they lost their flexibility and thereby tended to fracture abruptly, posing life safety issues. Plastics, even when reinforced, did not perform well under point loading, typical of conventional construction. To minimize point loading and provide a strong, rigid surface without excessive use of material (which would increase the overall weight of the structure), plastics needed a high modulus of elasticity. In the end, plastics were better suited to continuous curved compound structures due to the fact that they could be easily shaped and structurally required the benefits of compound curved forms to become viable structural building materials. In effect, the choice to use plastic governed the formal requirements of the architecture—at least, so it seemed in theory.

Non-compound curved forms were discussed by the MIT architects in a formal statement written prior to the development of the final design for MHOF but were not considered practical. Structural tent projects that could be easily mass-produced in kits and readily distributed to a variety of sites had some promise. It was determined, however, that they posed too great a risk, as they produced irresolvable technical problems of security, piercing, flutter, and stability in high winds, among others; they also created heating and maintenance problems considered insurmountable for the family's "space budget."[72] MHOF needed to be an economical "space age" house, sized appropriately for a growing family's income. With long-term growth potential in mind, the MIT architects also proposed a cellular house that could expand in modular increments with flat sandwich panel walls and dome roofs in honeycomb configuration. This typology, however, seemed unlikely to respond well to one essential question posed by the researchers: "Could the site come and go as it pleased under the house?" The team was interested in a modular house formulated as a kit of parts that could accommodate various site conditions. Although a honeycomb house might be designed to meet that criterion, it would not necessitate a new structural system specific to plastic: its flat wall surfaces would not adequately showcase plastic's potential.

The MIT architects agreed on a design in which "the compound curve might be the total enclosure." They favored the idea of "a continuous curving surface [that] could result in the floor extending to form the wall and finally the roof and ceiling." As they reported to Monsanto: "This was a concept that very few structural systems and materials are capable of accomplishing. The ideal form had to be one in which the enclosing material was of a continuous surface; the floor, walls, and roof all being of the same geometry."[73] This, in effect, was the form in which plastic might best exhibit its promise, unique from and exclusive of most other materials. If Monsanto could cultivate desire for this new "plastic aesthetic" along with the technical specifications to prove it was achievable, it might be able to ensure the long-term market success of its product.

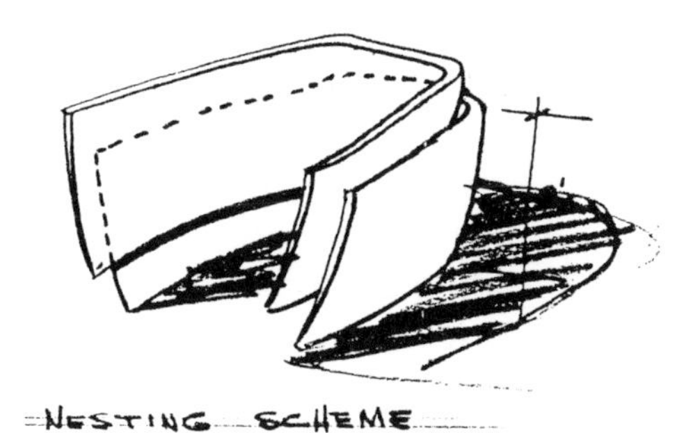

Architect's design sketches of continuous U-shaped bents for MHOF

Monsanto's Market Development Department, in conjunction with MIT, developed criteria for an ultra-modern house of the future. They maintained that it must be designed for "ultimate spatial quality and usefulness," "flexibility of size," "economical fabrication to enable the resulting building to be as large as possible," "ease of erection," "flexibility of siting," and "suitability in the landscape"; in addition, the design "would not hinder the individual's ability to reflect his personality."[74] In response to these criteria, MIT designed MHOF using modular "continuous" shells that formed roof, wall, and floor. These U-shaped shells with glass infill ends called "bents" were limited to eight feet by sixteen feet for transportability. Their curved surface and sandwich construction rendered them "capable of absorbing extreme shell stresses with a minimum amount of material."[75] These segments were designed to resemble an airplane wing cantilevered off a fuselage, where the symmetrical forces from each wing are carried indeterminately and continuously through the central core. In the article "Engineering the Plastics 'House of the Future,'" Dietz explains that this idea was borrowed from airplane designers, who carry the wing structure straight through the fuselage and avoid putting wing stresses into the fuselage. "The wing is one big unit."[76] In MHOF, "the floor and roof sections were [similarly] designed as units so that the load is carried straight through and not transmitted to the columns to any great extent except for the vertical load."[77]

The structural U-shaped shell "wings" were to connect together through a square central mechanical and plumbing core. The core housed the kitchen, bath, and circulation systems, while the dining room, living room, and bedrooms were fully enveloped within the wings. In essence, the design concept was to inhabit the space between the compound shell curves of an airplane wing, designed as a lightweight kit of parts to minimize construction time and materials in a form modulated to the habits of everyday life. Wartime advancements in easy-to-install, mass-producible, unbreakable, lightweight, waterproof, continuously molded plastics were transmuted both conceptually and aesthetically into a new, modular, domestic spatial tectonic.

Like a fiberglass plastic pool, MHOF could be put together by fathers and sons in their own backyard. The composite plastic bathroom core was built as one continuous lightweight element that came fully equipped and ready to install. Inside the walls the U-shaped bents were what the MIT architects described as a flexible "system of sub-frames": "The man of the family could then demonstrate his do-it-yourself creativity by designing within the structural framework of the shelter the eight window walls of the basic house."[78]

Expandability was achieved through the addition of components, and flexibility was ensured by multiple layout possibilities. Although typically featured on flat terrain, options included houses suspended above the ground on concrete compression cores. As in Buckminster Fuller's Dymaxion House, mechanical and electrical systems were centralized and located in the core. For MHOF, this created a zone in which all heavy equipment could be installed structurally independent of the plastic bents.[79] The architects suggested that "by adding a second...foundation core, an increasing number of cantilevered wings could be added to the original structure." These expanded versions of the house "could reflect the added leisure time available to the family [by] providing do-it-yourself hobby rooms, TV areas, sewing rooms, etc."[80] MHOF was designed with the premise in mind that it could go anywhere and expand infinitely, on a site that might "come and go as it pleased under the house"; the structure "might in its entirety be lifted off the ground if the site so required."[81] The cantilevered wings were curved to provide stiffness against buckling and sloped to match changing bending moments. They formed "strong, light, and stiff box-shell monocoque structures [while] at the same time they were the expression of the structural function of the wings."[82] They were a symbol of their own structural integrity, held up on a concrete plinth as a representation of the promise of plastic form.

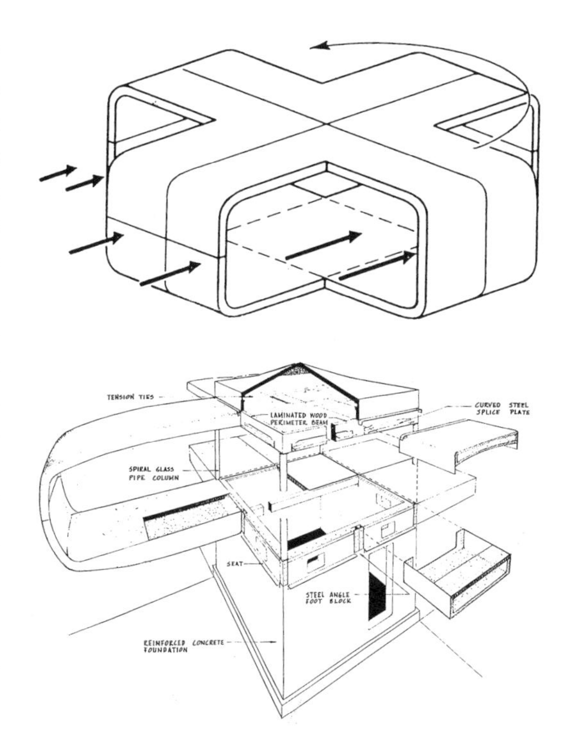

MHOF, structure and wind load diagram (top) and box construction diagram (bottom)

In the execution of the MIT design, the architects and engineers faced numerous problems. The continuous U-shaped bents eventually had to be discarded for practical reasons concerning their structure, manufacture, and distribution. A joint was established between a lower and upper bent, allowing for more reasonably sized modules, better structural independence for each bent, and greater integrity against thermal effect and wind loads.[83] The stiffness of the plastic materials had to be significantly increased, and according to Dietz, "to get this stiffness, the sections had to be made deep enough to have a large moment of inertia without using a tremendous quantity of expensive reinforced plastics."[84] In effect, the material resisted their ideal conception of "plasticity."

To solve all these problems of structural force, a box girder system was adopted, providing a structural frame above the concrete core to which the cantilevered wings, roof, and floor were then attached. The floor and roof were designed as large, hollow girders, and four laminated-wood spandrel beams were added to the roof at the core to transfer the loads. The floor was made independent from the bents as a sandwich panel supported over a wood beam, and the ceiling had to be soffited. Reinforced plastic ribs were added as needed to increase stiffness. Reinforced columns were designed at the four corners of the core to support the roof and sunk down into a reinforced concrete foundation; they were then stiffened in the foundation area by large gussets.[85] In the end, some of the interior partition walls also needed to be used as permanent structural panels to transfer lateral forces. A series of steel bolts, connectors, and machine screws were used in combination with adhesives to attach the modules together, then smoothed over to simulate "continuity." In effect, MHOF was framed as a box with cantilevered "curved" plastic floor and roof elements that needed extensive conventional reinforcement beyond compound curve "plasticity" to achieve its physical form. It was not the embodiment of an idealized engineering of plastics' "inherent properties" but instead the demonstration of an effort to achieve *monocoque* form through fairly conventional technologies of the time using composite plastics.

Although MHOF used many traditional construction technologies in combination with plastics, numerous tests were still necessary during the design phase to ensure and prove the strength of the proposed structural systems. Test bents were constructed at the Monsanto plant in Springfield, Massachusetts. They were subjected to strength tests using barrels filled with water to simulate the weight of 150 people packed into each room or about 5 feet of snow on the roof.[86] Thermal tests were also conducted using oscillating sprinklers that exposed the bents to 186-degree temperatures. The ultimate test of MHOF, however, was then carried out through the exhibition of the full-scale prototype.

In the August/September 1956 issue of *Monsanto Magazine*, images of the architect's plastic design model of MHOF appeared

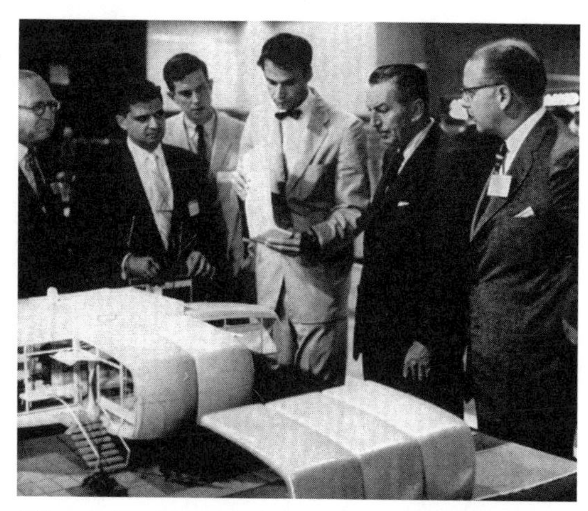
MIT researcher showing plastic "kit of parts" to Walt Disney, 1956

along with the question, "When can we expect this project to be brought down out of the clouds and planted for people to really see and believe?"[87] This visionary project was scheduled to descend to earth the following year. However, at a price of $1 million, Monsanto soon realized that their prototype for the House of the Future would not likely be mass-produced for the market well before it was even built. It could only be "a show piece designed to show the way to greater use of plastics."[88] Mueller announced to the company in October 1955 that even the basic plastic materials for a 30,000 pound house, at 50 cents per pound, would start at a cost of $15,000—already well beyond most peoples' budgets at the time—and that the actual cost of the finished house was indeterminate. MHOF was to be a "demonstration and test house," exhibited where it might have greatest effect. "The encouragement which this house would give to various people interested in plastics," Mueller maintained, "would be sufficient to push along the use of plastics in housing."[89]

Disneyland, in Anaheim, California, was selected as the site for the MHOF demonstration house, based on the location's potential to test the market and gain the greatest amount of publicity for the project. Monsanto had already established a successful relationship with Walt Disney. Monsanto's Hall of Chemistry had been in operation at the amusement park from the time of Disneyland's founding in 1955, and Monsanto believed the exhibit successful in demonstrating the newest in chemical production. The company had large investments in California, due not only to the state's proximity to natural resources but also to its growing aerospace industry. Cognizant of the fact that Disneyland, located "in the heart of the great Southern California population area,"[90] attracted visitors from every state,

Monsanto executives set out to "capitalize on the flow of people and their reactions over... years of exposure."[91] MHOF was designed to be tested by millions of people.

Disney conceived his theme park as a place that would incite nostalgia for the past and childlike adventurism of the future. "Here you leave today and enter the world of yesterday, tomorrow, and fantasy," declared Disneyland's 1955 dedication plaque. Incorporating their experience from animated film, Disney and his team of "Imagineers" built a fantasy-filled, consumerist entertainment mall based on studies of a wide variety of successful world's fairs, theme parks, and urban centers. Disney tapped into an unfulfilled longing in Americans to escape from the harsh realities of urban life and to spend the day with the entire family in a miniaturized, walkable city full of entertainment and adventure.

The fantastic structures that comprised the new urban theme park were constructed primarily of complex steel and wood frames covered with wood lath and plaster sheathing. As the *California Plasterer Journal* noted in July 1955, "Only these products could successfully answer the call for materials to fit the many shapes to be expected of a Walt Disney design enterprise."[92] Fantasyland Castle and Snow Mountain (later replaced by the Matterhorn) were not shaped of stone or earth. What a building or theme event looked like had nothing to do with how it was built or what materials were used in its construction.

Many of the themed areas inside the park—Main Street, Fantasyland, and Frontierland, for example—were based on historic neighborhoods presented on television or in the movies. Tomorrowland, however, where Monsanto leased exhibition space, was created to present the future and proved to be the most challenging area of the park to conceive and construct. When it opened in 1955, little more than a bunch of balloons hid the lack of invention demonstrated by a series of dressed-up storage sheds that comprised the "carnival rides" of Tomorrowland.[93] Disney had originally intended his futuristic city to showcase innovations in technology and industry. His Imagineers had dreamed up monorails and rockets

MHOF, structural bents, set in place on the Disneyland site, 1957

streaming throughout a new utopian city raised off the ground on stilts. None of these images, however, were as yet the least bit constructible. Anxiously awaiting the arrival of new technology, Disney welcomed and supported Monsanto's intent to design a Fiberglas plastic House of the Future—especially one that might provide the technology for an inexpensive, malleable building typology well-suited to space-age fantasy.

Late in 1956, Walt Disney contacted Imagineer John Hench to coordinate with Monsanto and MIT on their housing plan. Walt Disney sited MHOF just outside the entrance to Tomorrowland marking the important transition from the Fantasyland Castle. Inside the boy's bedroom, visitors would have a clear view of Snow Mountain.[94] MHOF would be the first significant attraction at Tomorrowland, which held the potential to fulfill the utopian dream of a new domestic architecture.

The house set down in Disneyland in June 1957. Premanufactured and shipped from the Winner Manufacturing Company in Trenton, New Jersey, preparations for construction had taken three months. The shells proved to be a challenge to build, as ensuring continuity at the connections required extreme accuracy. Each bent was formed by first building a precisely scaled wooden mold of the desired shape. A negative mold was taken from the original using polyester resin and Fiberglas cloth. The negative mold was then used as a surface to build up an actual GRP bent using hand lay-up techniques. Ten piles of woven Fiberglas mat were layered between polyester resins and, once cured, each shell was insulated and stiffened with rigid-polyurethene foam, sprayed on the inside. Another, thinner interior layer of GRP was applied by hand and was cured at room temperature.[95] The structural bents were installed with a crane and finished on the site—hand-trimmed, sanded, epoxied, bolted, and riveted together. The assembly process took a total of three weeks. The process was extremely low-tech and labor intensive. The panels for floors, ceilings, interiors, and windows all amounted to more pieces than intended. In the end, the actuality of production and construction contradicted the image of the sleek, time-and-materials-saving, modern, plastic kit of parts. MHOF was not manufactured at a push of a button.

Nevertheless, image is sometimes all that matters, and ultimately the house proved to satisfy "space age" fantasies of the future. A reporter for *The New York Times*, in an article titled "Four Wings Flow from a Central Axis in All-Plastic 'House of Tomorrow,'" noted that, due to its curved white walls, the house "with the drapes drawn [gave] one a feeling of being in the cabin of a rocket ship headed for Mars."[96] Due to its small size, the reporter considered MHOF somewhat "claustrophobic" for the man of the house, but for the woman, it was considered a dream come true. Everything could be operated from a central command center: "The kitchen sink in this arrangement becomes practically a control tower, where she can maintain surveillance over three of the four rooms in the house. By pushing an array of buttons she can regulate practically everything but her husband."[97] There was even a button in this new space capsule equipped to control the air, with different scents for each room depending on changing environmental factors. MHOF had all the gadgets of a fantastic science exhibit. The house of the future came with all the promise of hi-tech automation. The kitchen was outfitted with three refrigerators (one specifically for irradiated produce), a range, and laundry equipment. All of these appliances were able to slide within walls, drop beneath counters, or raise and lower from the ceiling. The kitchen was designed to disguise its appearance "because it is fully viewed from the dining room and the living room." As for the furnishings, the most up-to-date comforts were distributed throughout the home. From tables and chairs to sliding panels, rugs, fabrics, cups, plates, countertops, pillows, drapes, sinks, toys, phones, and dolls, everything needed for a comfortable home-life was included—and manufactured in plastic. Even if consumers could not yet purchase a Monsanto House of the Future, they could surround themselves in furnishings made of this "revolutionary" material.

MHOF, construction near completion, 1957

Encased in the comforts of a modern home, the project provided a sense of security that might pacify latent fears in a Cold War society. Exhibited next to the castle of Sleeping Beauty, this house projected the fantasy image of a space lander or mobile bomb shelter. The house provided for domesticity within a (theoretically) transportable kit of parts, tailored to a future lifestyle of speed and push-button efficiency. MHOF advertised future consumption as the means to safe, modern living under the guise of a theme park tourist attraction. MHOF was an exhibition house for industry that masqueraded as an amusement ride, selling tickets to incite desire. And it was an extreme marketing success.

Over 435,000 people visited the Monsanto House of the Future within the first six weeks of its grand opening. Media coverage was extensive and included everything from local television spots to articles in *The New York Times* and *Time* magazine. Monsanto estimated that 5 million people visited the house per year, which translated into "10 million footsteps in 12 months" testing the vinyl tile floor.[98] In 1957 Monsanto estimated further that after "two years of 'scientific farming' by a staff of public relation specialists from 13 companies . . . a half-billion readers, viewers and listeners . . . [had] been exposed to stories of Monsanto's House of the Future."[99] In 1958 Disney's Consumer Relations Division took an exit poll for Monsanto from visitors to MHOF.[100] Of the 1,008 people surveyed, over 95 percent found the house to be entertaining, educational, or both. The kitchen and bathroom were considered by far the most

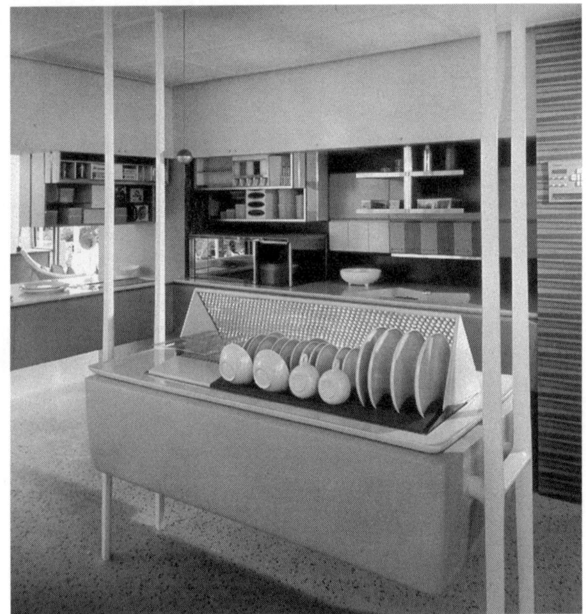

MHOF, 1957, dining room (top); living room (bottom, left); and kitchen (bottom, right)

interesting features of the house. The greatest complaints were that MHOF was too small, cold, and not very homelike. Nevertheless, over 96.9 percent of those surveyed truly enjoyed their visit to the House of the Future, even if only slightly less than 63 percent of them actually knew that Monsanto had sponsored the attraction.

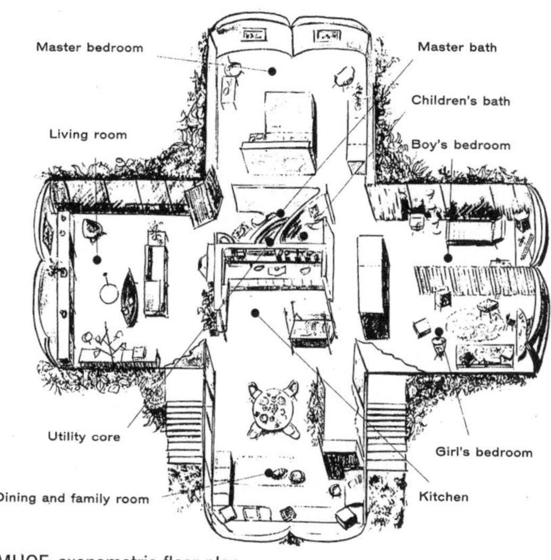

MHOF, axonometric floor plan

Monsanto's efforts to get the most from their investment continued for years. The house was renovated twice, with all new interiors created to display more homelike styles in contemporary (plastic) living. In 1958 a replica of the plastic shell traveled to the Brussels World's Fair and was exhibited in the "Face of America" pavilion.[101] According to Monsanto, in 1962, Russia announced that it had "achieved a construction breakthrough and built the 'first plastic house,'" a photograph of which they claimed bore "more than just a casual resemblance" to the House of the Future at Disneyland.[102] Monsanto was delighted by the success of its project, as compared to the Russians; MHOF had gained significantly more attention than its foreign counterpart. Of course, neither the Russians nor the Americans can be credited with the first "plastic" house; the French had already claimed that honor back in 1955 with an exhibition house by Ionel Schein.

Despite the renovations and publicity, by 1968, MHOF, and Tomorrowland in general, were no longer futuristic enough. Walt Disney decided to update Tomorrowland and, with the help of Charles Allen Thompson from Monsanto, MHOF and the Halls of Chemistry were replaced by another Monsanto attraction, "Adventures thru Inner Space," which featured a multi-media projection journey into the "World of Molecules and Atoms."[103]

Visitors test the MHOF prototype at Disneyland, 1957

MHOF was dismantled after over twenty million visitors had ventured on board. The demolition crews, however, were "baffled when [a] 3,000 pound steel headache ball simply bounced off [its] plastic walls."[104] Workmen were challenged to demolish the project, employing "torches, chainsaws, jackhammers, clam shovels and virtually every tool in their armament, to no avail."[105] An article in *Monsanto Magazine* described:

> Eventually choker cables were used literally to squeeze the big plastic modules into pieces small enough to be trucked away. Attempting to dislodge the house from its concrete pad, the wreckers found that the half-inch steel anchor bolts broke before the glass fiber-reinforced polyester material.[106]

In their very efforts to prove plastics could be structurally durable, the architects and engineers hired by Monsanto designed a modular kit of parts that could hardly be disassembled. All good intentions behind a flexible, interchangeable, mobile, and transportable architecture inspiring the birth of a new, second-order modernism were ultimately undermined by the very insistence upon durability, permanence, and fixed-stable form.

The plastics industry had its own agenda: to prove its usefulness in a building economy dominated by steel. Toward that goal, it may have proved plastics to be extremely strong, but it hardly proved them to be cost-effective. The MIT architects were disappointed that they were never able to mass-produce the Monsanto House of the Future. They came to recognize that national and state codes were not going to accommodate plastic as a structural material. As Dietz understood, there were still no industry standards for plastic structural materials, because structural plastics had not been around long enough to be tested for long-term viability against weathering elements over the course of a building's lifetime.[107]

Demolition of Monsanto's House of the Future, 1967

Plastics continued to find use in everyday domestic products and in the building, shipping, and aerospace industries, but they never garnered acceptance on the scale Monsanto and MIT had envisioned. Even in the 1970s, when fiberglass plastics found their true calling at Disney World, they were mostly used throughout the theme park as skins to sheath frame construction. Compared to much of Disney architecture, MHOF was significant in that it was not a frame structure wrapped with a free-for-all skin. The skin had a relationship to its structure—it was, in fact, structural. Its form was shaped to challenge the technology of the time and redirect the consumer housing industry away from frame construction toward alternative formal practices. MHOF was created to symbolize the freedom "plasticity" might one day provide the building industry, but in the end its structural integrity proved to resist the rhetoric of its fantasy form. Monsanto played an historic role in marketing the desire for "plasticity" and the promise of tension-shell construction in the 1950s but, despite its better efforts, was never able to overcome plastics' own "innate"

In breaking away from the city, the part became a substitute for the whole, even as a single phase of life, that of childhood, became the pattern for all the seven ages of man. As leisure generally increased, play became the serious business of life; and the golf course, the country club, the swimming pool, and the cocktail party became the frivolous counterfeits of a more varied and significant life. Thus in reacting against the disadvantages of the crowded city, the suburb itself became an overspecialized community, more and more committed to relaxation and play as ends in themselves. Compulsive play fast became the acceptable alternative to compulsive work: with small gain either in freedom or vital stimulus.

—Lewis Mumford

Toward the end of 1950, President Harry S. Truman hosted the Mid-Century White House Conference on Children and Youth. The conference brought together well-known authorities to deliver papers on various aspects of childhood. Social and educational problems, economic issues, and plans for implementing specific government programs aimed at children were all addressed. Of particular interest was the paper entitled "What We Know about the Development of Healthy Personalities in Children," delivered by Dr. Benjamin Spock, the established authority on children and parenting of the time. His description of the developing pre-schooler states:

> The boy becomes increasingly aware that his destiny is to become manly, in the pattern of his father and other admired males. He plays at driving cars, shooting guns, building skyscrapers, going to work. The girl who is devoted

> to her mother takes joy in turning more and more to doll care and other feminine fascinations.[1]

Spock's was the voice of modern American child culture, and his statement epitomizes the mindset of the engendered society prescribed in the postwar period. According to the doctor, the boy has a destiny that can be articulated in detail, whereas any "feminine fascination" other than "doll care" is not worthy of description. The boy's destiny is a kind of manliness derived from the image he holds of his father. Each aspect of this gender-specific model for adult occupations is ultimately reducible to a corresponding form of child's play: driving cars—as in a race car driver; shooting guns—like a war hero, a western cowboy, or a pioneer settler; building skyscrapers—like a construction worker, architect, or engineer; and finally, the generalized notion of "going to work"—still a vague concept to the boy, as he never actually witnesses his father in the workplace but has only a notion that some activity happens in his father's life while he is "at work" and not at home. According to Dr. Spock, boys are little men and the lives of both are defined by modes of play in accordance with the reciprocal model of work/play he advocates. Girls, as the protégés of their mothers, simply find joy in doll care, just as their mothers do in childcare. In addition to affirming idealized, gender-specific roles for children and adults in suburban society, this framework echoes a perceived *de facto* correlation between childhood and adulthood that pervaded all aspects of the suburban domestic realm in the post-World War II era.

Emulation and mimicry reduce adult work activities to the level of child's play, a reflection of underlying processes described in labor-play theories that posit the transformation of certain modes of work into play. According to the philosopher and political theorist Hannah Arendt in her 1958 *The Human Condition*, "every activity which is not necessary either for the life of the individual or for the life process of society is subsumed under playfulness."[2] By first clearly delineating the boundary separating play from necessary human activity (i.e. labor), and then positing the category of work against the category of labor, Arendt distinguishes one part of the human experience from the other. She explains, "The emancipation of labor has not resulted in an equality of this activity with the other activities of the *vita activa*, but in its almost undisputed predominance. From the standpoint of 'making a living,' every activity unconnected with labor becomes a 'hobby.'"[3] This notion of hobby reinforces the central position of play and the incremental transformation of various modes of work into play in postwar American culture.

In part, the emancipation of labor described by Arendt is accomplished with the aid of specific technological advances that break down work into a series of automated tasks. Machinery, appliances, and other sophisticated mechanical gadgets erode the work process by transforming it into an array of button-pushing exercises. By reducing the requisite number of physical and mental operations, technological expedience and convenience reduce the workload, freeing up time for leisure and play, while the increased automation of the work process itself conforms more directly to modes of play.

A burgeoning middle class grew out of a cultural context that privileged leisure time over work. Eventually, the liberation from work became one of the central ideals of the American, suburban, middle-class lifestyle. Miniaturized toys such as scaled-down versions of kitchen appliances, lawnmowers, doctor's kits, and chemistry sets—many of them functional and not simply representational—although not innovations of the postwar toy industry, were marketed and embraced as preparatory devices for training the next generation of American children for their gender-specific future roles as citizens of the United States.

The relationship between the toy and the full-scale appliance that linked the plaything to the machine equated child's play with grown-up labor. It also had the reciprocal effect of diminishing adult work into a form of child's play. As the principle of leisure became increasingly embedded in the postwar American lifestyle, suburbanites embarked on a constant quest for more free time. Housework and other domestic chores were reduced to a series of timesaving activities. The task of preparing a meal or tending to the front lawn

The Easy-Bake Oven, manufactured by Kenner Products, 1962

no longer involved lengthy and arduous processes but were simplified and expedited with the aid of tools such as instant ingredients and automated machinery. This contributed to the blurring of boundaries between work and play in the domestic realm. The American market targeted women by advertising pre-packaged TV dinners and instant or powdered food products as extricating the modern housewife from the so-called burdens of domestic life. Soon, nearly every consumer product was "quick & easy" or "time-saving" and was accompanied in the marketplace by a variety of appliances and gadgets designed to further lighten the load.

In her critique of the claims of American sociologists titled "For the Sake of the Children," the social critic Anne Kelly explores the shift in opinion prevalent among specialists in the late 1950s concerning childhood and suburban life. She questions the reassessment of suburbia as having a negative impact on children by sociologists who once considered it beneficial to child development. She asks if some of the drawbacks of postwar suburbia, rather than being inherent to suburban life, might not reflect the broader shifts in American culture toward conformity and conservatism.[4] Essentially her argument supported the notion that if the move to the suburbs was ultimately for the sake of the children, as was often claimed, then the subordination in some way of every aspect of suburban life to the needs of the child was not a surprising end result.

A decade later in his 1969 *The Suburban Myth*, the journalist and American cultural critic Scott Donaldson rearticulates the fundamental claim made by Kelly:

> It is for [the children] we moved here in the first place. It is for them we battle at the village hall in terms of more schoolhouses and playgrounds. They're the reason about a third of our shops are in business and the milkman stays employed. They're responsible for a good part of our reading in books, magazines, and newspaper columns. They explain our having a television set we vowed we'd never buy. Suburban culture is child-centered.[5]

If we accept Lewis Mumford's eloquent characterization of childhood as "the pattern for all of the seven ages of man" and examine it in conjunction with Dr. Spock's prescribed roles for postwar women, men, boys, and girls, it becomes increasingly difficult to differentiate between the children and the child-like adults who populate the suburban realm. The German cultural philosopher and specialist on youth and children Dieter Lenzen observes in his 1989 critique of the postwar expansion of childhood, which continues to this day, "It is not childhood that is disappearing, but the status of the adult—brought about by an expansion of childlike aspects in all spheres of our culture."[6]

The post-World War II increase in leisure time demanded a corresponding increase of leisure space. The initial signs of the infiltration of leisure activities into the domestic realm can be detected in the shifting design standards for the suburban home. Domestic architectural design of the period responded to these demands by expanding on innovative design strategies first introduced in the 1930s. Well before the postwar suburban explosion in the United States, the architectural firm of Landefeld & Hatch produced a 1,100-square-foot housing prototype that incorporated a recreation room as a central feature of its design. Out of the fifteen demonstration houses which constituted the "Town of Tomorrow" exhibition at the 1939 World's Fair in Flushing Meadows, Queens, the visiting public voted Landefeld & Hatch's the most popular.[7] A decade later, the inclusion in residential design of a space dedicated solely to play would become commonplace.

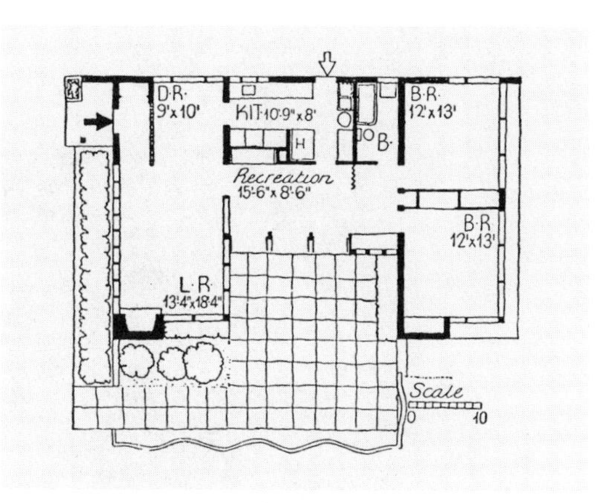

Landefeld & Hatch Architects, World's Fair House, 1939, with its large central space dedicated to recreation

In 1950 the editors of *Architectural Forum* attributed the desirability of a separate room devoted to child's play within the suburban home to the basic structure of the American family:

> As children grow into adolescents, their hours of leisure tend to coincide with their parents' and there is competition for use of the living room. The tired father returning from work will become an immovable object in the living room absorbing peace and quiet before the next day's demands have to be met. At the same time his adolescent children will blow into the same living room like some irresistible force—full of noisy physical exuberance to offset the confining routine of a school day.[8]

This influential professional journal offers evidence for how exemplary architectural design solutions in the domestic realm were predicated on individual family members' needs vis-à-vis leisure time and space.

William J. Levitt, House in Levittown, Long Island, N.Y., 1947–48, interior

Only the most innovative suburban homes of the early postwar period possessed both a living room for formal occasions and a family room large enough to accommodate games, the TV, an indoor barbecue, and general clutter.[9] However, close examination of a typical living room in one of the six thousand modest Cape Cod houses built by William J. Levitt in Levittown, Long Island, between 1947 and 1948 reveals the impact that emerging spatial needs coupled with popular cultural assumptions about family life had on the suburban plan. In this living room, each member of the household is allocated space, yet each of these spaces intrudes upon the others. All possess a sense of mutability and essential utility as a space for play. The card table with accompanying wall ornaments—a sort of "dad's corner"—is situated in opposition to the "mom's corner," where various laundry accessories and a bounty of up-to-date homemaking magazines delimit a feminine zone. Nearby, neatly tucked into the wall, is a bar for large social gatherings with friends and neighbors. The strategy of spatial organization leaves the majority of space open for the entire family to occupy, however and whenever they are at play. These conditions precipitated the inclusion of both a formal living room and a play room in the suburban home of the 1950s.

The study of spatial dynamics in the postwar suburban home became a serious concern for designers, as demonstrated by the planning models and usage diagrams developed by the architect-builder Haydn Phillips in 1950.[10] Not by chance Phillips's work appeared in the April edition of that year's *Architectural Forum*, the issue dedicated each year to housing. The editors cited the record of $4.4 billion spent on new construction in the first quarter of 1950, out of which $1.9 billion went to new home building. Many of these homes, which were relatively small in size, were built without the help of an architect, as had been the majority of homes built in the previous year:

> Merchant builders erected perhaps five out of every six single family homes last year—a staggering total of nearly 700,000. Architects, unfortunately for themselves, for the builders and for the home-buying public are playing no such part. Of homes costing $12,000 or less, perhaps one in three had the direct benefit of an architect's skill, imagination and training.[11]

In light of these statistics, *Architectural Forum* dedicated the issue to the theme of "The Small House" and to encouraging "a better understanding between architect and builder."[12] The editors of the magazine sought to promote better cooperation between architects and builders by advocating quality home design through an improved understanding of the needs of the client. A variety of residential projects included in this issue synthesized superior design solutions with cost-effective building practices. In an introductory message to the reader, Thomas P. Coogan, president of the National Association of Home Builders, wrote, "The builder on his part must realize that good design is the one thing every home buyer wants, whether he realizes it or not."[13] Surprisingly, even the foremost representative of the home-building industry had determined that many homes built without the creative assistance of architects failed to address the needs and desires of the suburban family.

The article "Quality Houses through Contemporary Design" endorsed seven design strategies that were both cost-effective and

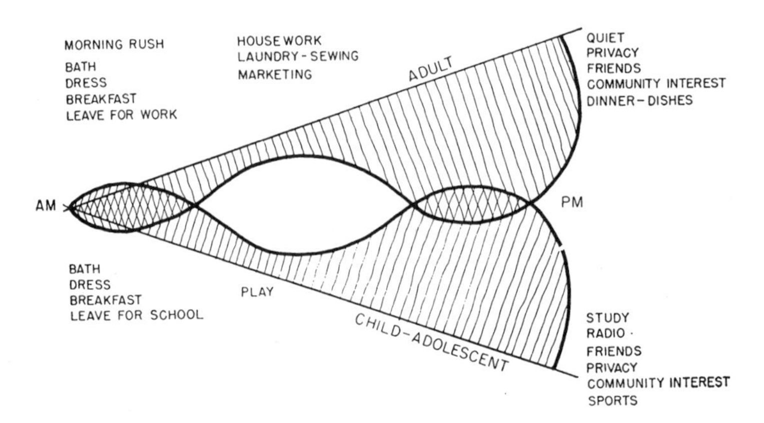

Haydn Phillips, graph of living activities, 1950

would improve "the allocation of space to the various phases of family living and the relationship between these areas":[14]

- orientation of the house to the view, sun, breeze, and other climatic conditions
- openness of areas
- convenience of household operation, so that, for example, "the kitchen becomes the control center from which children's activities may be supervised"[15]
- flexibility of space use, which is described as "going hand in hand with convenient open planning"[16]
- expandability
- integration of indoors and outdoors, "aimed at making the house and its surroundings work together and for each other[17]
- adequacy of the storage space for household effects

Phillips's spatial analysis was included in "Planning for Complete Flexibility." This article also featured six of his prototypical floor plans along with diagrams that described space use in order to emphasize "the possibility of maintaining a shell of exterior walls during a whole generation of family life and adapting the interior to satisfy the crescendo of children's needs within the shell."[18] These needs centered around play and leisure activities and were juxtaposed against the way adults occupy and use the domestic space. In his graph of "living activities," Phillips created a spatio-temporal diagram that delineates when and where the lives of adults intersect with the activities of children. His diagram positioned play against housework, laundry, sewing, and marketing as two diverging vectors describing daytime activity within the domestic space. He charted two oppositional curves that represent the daily spatial demands in the suburban home.

Exactly how Phillips determined the slope of his vectors and by what standards he was able to gauge the magnitude of the specific activities that serve as his criteria for spatial demands is unclear. Significantly, however, he claimed that the so-called "space problem" in the suburban home was temporarily alleviated around noontime. It is precisely at this location along the child-adolescent vector that he located the term "play." The prominence of the term (despite its vague application) in the graph confirms that play was seen as an integral and dominant factor shaping suburban life in 1950. Ultimately, what Phillips's graph demonstrates is a fundamental primacy of children and their spatial requirements in postwar American domestic space.[19]

Adhering to the conclusions presented by Phillips, the article maintained that "in the early years, the children's nursery can literally be merged with the kitchen to allow big play areas within sight and call of the housewife. At night [a] partition will be pulled out to separate the noise of dinner dishwashing from the sleeping young."[20] This model for good design predicated on the woman's role in the household economizes on square footage by combining distinct usage zones into one common space and reaffirms the design strategy of flexibility of space use presented elsewhere in the *Forum* issue.

A built example of the kitchen-nursery is presented in the article "Builder Operations," featuring suburban tract houses designed and constructed by builders. Of the nine projects shown, the ranch-style house built by developer Gordon Bronson in central New Jersey was offered as a prime example of the successful combination of architectural and construction concerns and as an improvement on the

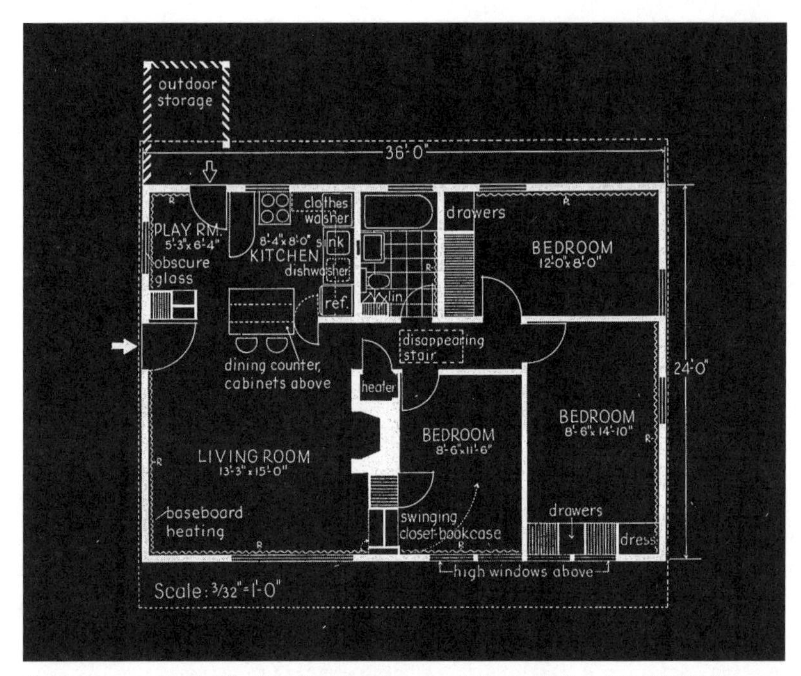

Gordon Bronson Construction Co., Three-bedroom Ranch House, Raritan, N.J., 1949–50, plan

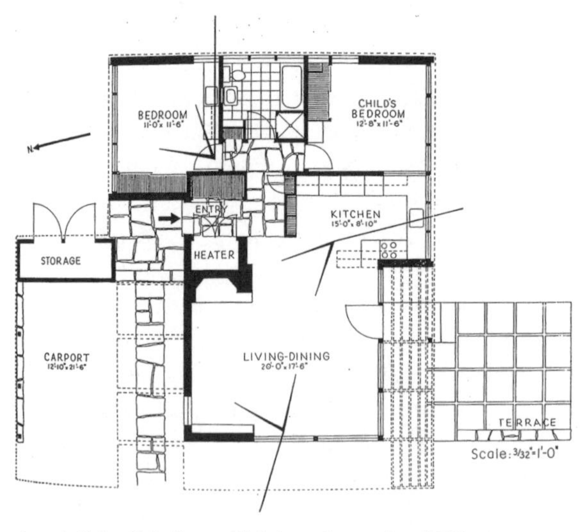

Joseph Stein, Stein House, Waterbury, Conn., circa 1950, plan

suburban paradigm established by Levitt. Whereas the designated living room space of the typical Levittown, Long Island, house was transformed by its owners into a multi-purposed play room to accommodate the needs of all members of the family, Bronson's prototype for a three-bedroom residence signals the formal introduction of the playroom as separate from the living room in low-cost suburban tract housing. Essentially part of the kitchen—which itself is joined to the living room—the playroom was presented as a natural and necessary element in the prototypical ranch house and was described by the editorial staff as fulfilling the need for both convenience of household operation and flexibility of space use.

The Joseph Stein House in Waterbury, Connecticut, directly incorporated all seven strategies for good housing design enumerated in the introductory essay of the magazine. The innovative design of this modest suburban home, built by the young architect to accommodate the specific needs of his own family, shows the influence of particular conceptions of domesticity from the postwar period.

In line with the notion that the suburban home was child-centered and that the housewife would need to constantly supervise the children, Stein chose an open plan that joined the kitchen with the living room in one contiguous space. For the "convenience of household operation," he added an interior window in the kitchen, opening onto the adjacent child's room, which enabled mother to maintain constant surveillance of her young daughter; in the words of the editors, Stein "followed [the] trend in design for family living: the pattern of planning so that children should play within the mother's view, having the run of the house."[21] Child's play had served as the primary generating force that gave the domestic realm its particular shape.

The final section of *Architectural Forum's* April 1950 issue in which the Joseph Stein House was featured, "Tailor-made Houses," acted as a counterpoint to "Builder Operations." It was designed to show "builders some of the best design ideas the best architects are now contributing to low cost housing, both custom and builder built."[22] The underlying editorial conceit of *Architectural Forum* is most clearly conveyed by the magazine's subtitle: "The Magazine of

TAC, Catheron House, Foxboro, Mass., 1949–50, interior

Building." From this vantage point, the *Forum*'s 1950 housing issue endorsed the increased cooperation between the trade (builders and developers) and the profession (architects and designers) without privileging one group over the other.

The topic of small house design was taken up a month later in the May 1950 edition of *Architectural Record*, which posed on its cover the question, "Do small houses afford satisfactory practice?"[23] Overall, the twenty-five houses shown in the issue answered the question in the affirmative. As if in an attempt to distinguish their issue dedicated to housing from the previous month's *Architectural Forum*, the editors of *Architectural Record* focused almost entirely on the work of architects and limited their presentation of houses built by developers. Although both magazines examine small and relatively inexpensive houses, *Record* is decidedly geared toward the professional audience, emphasizing its editorial position that the role played by the building trades was secondary to good residential design.

The section of the May 1950 *Architectural Record* highlighting housing was titled "Building Types Study no. 161." The first of its three segments was devoted to four New England houses by TAC, The Architects Collaborative.[24] The second segment, titled "The Revere Quality House Program," presented sixteen projects as exemplifying "good speculative home design."[25] Almost all the houses featured in this segment had open plans or over-sized living-dining areas, and more than one-third included either a multi-use, recreation, or play room. The final segment, "Architects and Prefabrication," focused on the Acorn House, a housing prototype

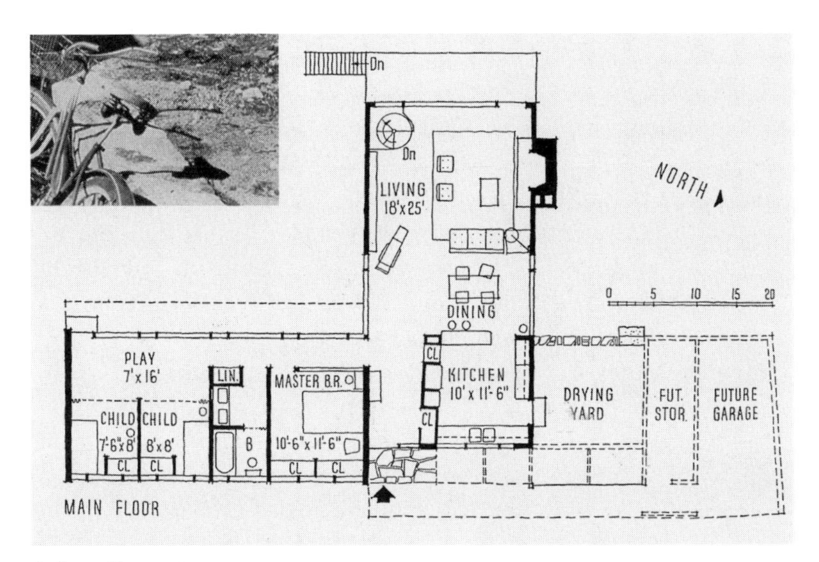

Catheron House, plan

developed by the architect Carl Koch and his engineering colleague John Bemis.

Of all the houses in this issue, the Catheron House in Foxboro, Massachusetts, received the most attention, garnering six full pages of images and drawings with accompanying text. That text described the unique approach of TAC to architecture and praised their collaborative working environment, commitment to good design, and relationship with both client and builder, as well as their innovative use of standard building materials in new and effective ways. The work of the Boston-based architectural firm—a self-professed "bossless" firm established in 1946 "in the belief that collaboration could be more useful in modern architecture than brilliant individualism"[26]—provided examples of houses that epitomize the highest ideals in residential design in the minds of *Record*'s editors. Four projects by TAC were featured in the issue; in each, the design of the domestic space is primarily achieved in response to the domination of child's play in suburban life.

In the Catheron House, the apparent separation between the common public space of the kitchen-dining-living room and the private wing containing bedrooms is deceptive. The two children's bedrooms at the end of the lateral axis are actually only alcoves with folding screens separating the beds from the hallway, which is designated as a common play area. By pushing the screens aside, the entire space is transformed into a single, expandable zone conducive to play. The architects' use of an L-shaped plan (rather than the rectangular footprint employed by builders such as Levitt and Bronson) and their glazing of most of the wall surfaces facing the yard creates a house that embraced the exterior space and effectively incorporates it as an additional room. Both the physical and visual connections established between the expandable play space of the children's quarters and the seemingly independent, common public zone of the kitchen-dining-living room extend the function of one area into the other, forming a unified whole.

In a follow-up article from June 1950, *Architectural Record* once again featured a unique housing project by TAC. Six Moon Hill in Lexington, Massachusetts, is characterized by the editors as an example of "collaborative planning [that] integrates tailor-made houses in co-op subdivision [and] demonstrates new ideas in design."[27] Expanding on the standards for good housing design presented in the previous issue, they endorsed the project on the following basis:

> As a group, the houses demonstrate that: 1) Through good site planning and harmony of design moderately priced modern houses of different shapes and sizes can be built to form a pleasantly coherent community, in sharp contrast to the monotony of most FHA-financed neighborhoods. 2) Owners of modern homes in a development safeguarded by this type of design control can get satisfactory financing from private sources. 3) With a little experience, contractors can put up a good modern house at low cost and a profit. 4) With professional control and a strong bond of common interest, cooperative housing ventures can avoid the pitfalls and technical difficulties through which many such undertakings have foundered.[28]

TAC, Stewart House, Lexington, Mass., 1949–50

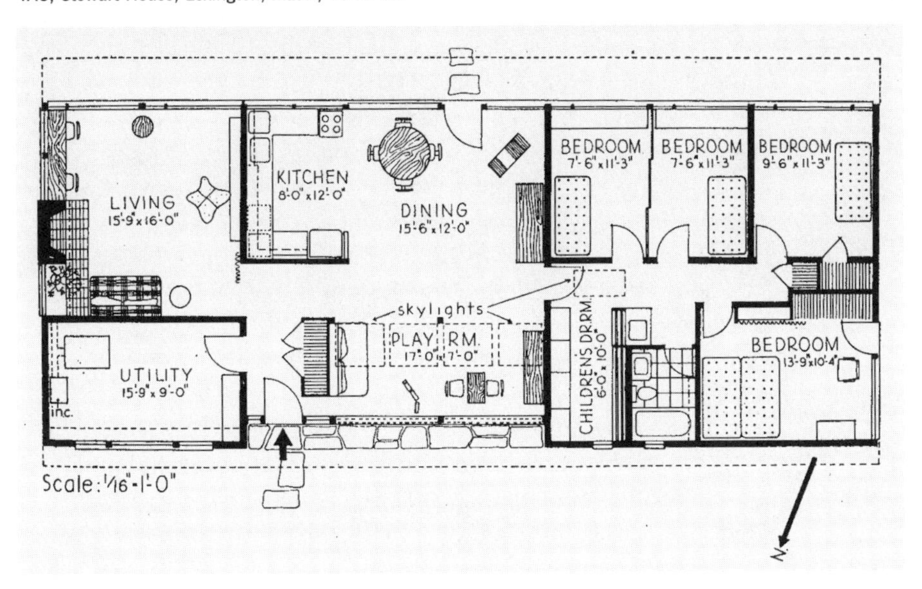

Stewart House, plan

In praising the housing complex, the editors vilify the type of developer/builder projects that conform to the basic Levittown model and advocate in the strongest of terms the increased participation of architects throughout the entire design and building process.

Of the eleven houses that constitute Six Moon Hill, the Stewart House provides the most explicit demonstration of the inundation of the domestic space by play. The playroom, having grown to the size of a formal living room, dominates the area adjacent to the kitchen-dining area, taking on new prominence within the overall interior space of the house. The kitchen becomes an auxiliary zone to the new focal point. The living room—now more like a living "nook" than the main social space of the house—is displaced off to one side as a satellite space. The architects of TAC completely inverted the traditional spatial hierarchies of residential design by abandoning the kind of formalism that placed the living room at the heart of the configuration. The zone designated for play is expanded through the use of industrial glazed garage doors that open up the entire length of the room to the backyard. The toys seem to spill out of the house and into the yard. No longer just visually contiguous—as in the Catheron House—the kitchen, dining, and playroom are here physically joined together and, with the backyard, constitute the largest "room" of the suburban dwelling. The well-designed domestic interior, whose configuration is culturally determined by the norms of behavior assigned to each member of the household according to gender and age, coupled with the dominant position of children within the hierarchy of spatial needs, now extends beyond the building envelope to fully encompass the backyard as well. By 1950, child-centric suburbia had produced explicitly child-centric architectural design.

No longer constrained by the idealized vision of the suburban home, with neat flower boxes and a white picket fence, postwar suburban housing took as its fundamental inspiration the increasing pervasiveness of mechanization along with the domination of play and child-centric design. The appearance of an expandable open space

Stewart House, interior

officially dedicated to leisure activities—in the most basic developer-designed, affordable housing to the high-designed housing provided by progressive architects—signals a fundamental shift in the paradigm of domestic design. Satisfying both the functional and ideological demands of the postwar period as they emerged in 1950, the product was a streamlined, efficient living environment that fully embraced the ideals of play and leisure. The emergence of the playroom as an autonomous and dominant element within the typical suburban home attests to the centrality of play among the concerns that were shaping domestic architecture.

In postwar America, play and leisure emerged as central motivational forces behind all modes of modern living. Suburbia and the suburban lifestyle embodied the idealized myth of small-town American wholesomeness and virtue combined with "modern" sensibilities and luxuries. All aspects of the suburban lifestyle—including the design and development of the suburban single-family house—focused around this ideal, and the slogan "for the sake of the children" became entrenched as the fundamental tenet of the period. "Suburbia" blossomed in the imagination of its advocates, whose numbers skyrocketed with the rapid proliferation of affordable suburban housing developments. Builders, developers, and architects found ways to give expression to the suburban idea. As the projects featured in the 1950 housing issues of both *Architectural Forum* and *Architectural Record* demonstrate, the needs of children and the central role of play as a space-determining criterion within the domestic environment that are foregrounded by the editorial stance of each journal reflect the postwar American attitude in which, as Mumford states, "play became the serious business of life."

TOY

TAMAR ZINGUER

Designing toys was very serious play for Charles and Ray Eames. On July 16, 1951, amid a plethora of political writings on world news and United States domestic problems, *Life* magazine published the article "Building Toy" featuring the first toy the couple designed that was actually manufactured. The article boldly displayed full-page photographs of Charles Eames assembling a tall structure with *The Toy* and of children playing inside different environments created with its triangular panels. *The Toy* was described as "one of the most imaginative playthings of the year," intended "to intrigue young men (5–10) who have an engineering or architectural bent and young ladies (same ages) with a homing instinct."[1] The instruction sheet

The Toy, designed by Charles and Ray Eames, as it appeared in *Life* magazine, July 16, 1951

proclaimed, "*The Toy* gives each one the means with which to express himself in big structures and brilliant color." Charles Eames added that it was made "for teenagers to decorate their rooms, and for parents to make sets for plays, pageants, ballets and parties."[2]

The architectural office of Charles and Ray Eames had designed children's furniture in laminated birch since 1945. They also designed molded plywood animals—horses, bears, frogs, and elephants—upon which children could sit or with which they could play. What started as an offshoot of their work with wood laminates evolved during the next fifteen years to include a series of objects relating to children: furniture, masks, costumes, and toys. They also produced films the main protagonists of which were puppets, toy buildings, and mechanical toys.[3] Some of the projects were intended specifically for children, while others, like the films, were addressed to all and conveyed the beauty and color, as well as the great sense of order, found in everyday acts of play.

In the opening of the film *Toccata for Toy Trains*, Charles Eames could be heard saying, "In a good old toy there is apt to be nothing self-conscious about the use of materials. What is wood is wood; what is tin is tin; and what is cast is beautifully cast. . . . It is possible that somewhere in all this is a clue to what sets the creative climate of any time, including our own."[4] Constructing with paper and creating large prismatic forms echoed contemporary spatial and formal concerns. By analyzing the manner by which *The Toy* delineated space and by uncovering the structures associated with its webbed forms, the creative building climate of the 1950s can be examined.

The Toy, designed and manufactured by Charles Eames (1907–1978) and Ray Kaiser Eames (1912–1988) in 1950, consisted of brightly colored triangular and square panels measuring 30 inches on a side.[5] Wooden dowels with pierced ends would slide through sleeves on the panels' edges, and pliable connectors (pipe cleaners) would join the stiffened forms to model a series of prisms, which, as the instruction sheet illustrated, could become a tent, a tower, or a theater set. The panels were made of a recently developed, plastic-coated,

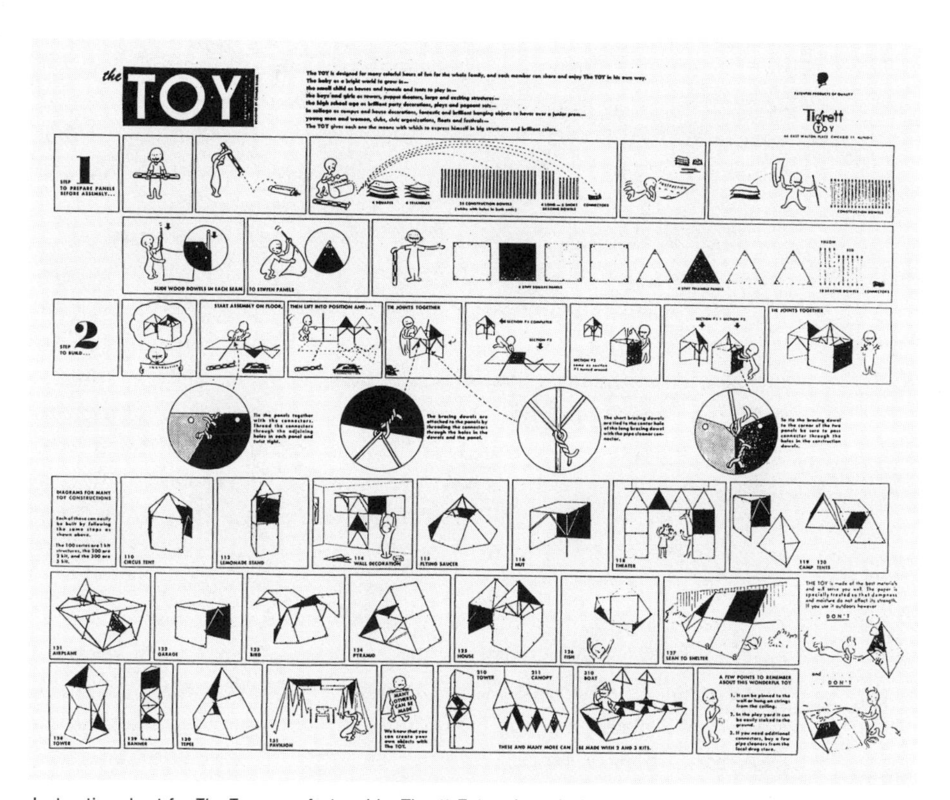

Instruction sheet for *The Toy*, manufactured by Tigrett Enterprises, Jackson, Tenn., 1951

Ray Eames, outside her home and studio in Pacific Palisades, Calif., playing with *The Toy*, August 1950

moisture-proof paper that was flexible enough to be rolled into a hexagonal cardboard tube. The writing on the tube proclaimed, "Large, Colorful, Easy to Assemble, For Creating a Light Bright Expandable World." "In a jiffy any child can have a real play-in-size house... or an airplane," a publicity leaflet promised.[6]

Charles and Ray Eames and their office staff built many three-dimensional and two-dimensional installations with *The Toy*. They constructed and then photographed numerous canopies, tents, and theater backdrops that mixed with the Eameses' furniture in creating colorful interiors. The grounds outside their office would fill with airy, open constructions conveying the impression of a light, collapsible, temporary world. In one photograph, Ray lies on the grass among three-dimensional, crystal-like shapes that look like space frames, towers, and structural totem poles. When configured as a two-dimensional form, *The Toy* became a decoration for the home, hung from the ceiling or spread out flat on the wall.

The Toy stood out among contemporary construction sets. Other building toys reflected the recent changes in the American landscape—the supremacy of the automobile and the development of the suburban home. *Fox Blox*, for example, a wooden toy manufactured in California in 1950, presented a system of slotted wood panels that assembled to form a ranch-type house, much like a prefabricated home. Another popular construction toy, *Plasticville, U.S.A.*, also first manufactured in 1950, allowed for the speedy erection of an entire town. Miniature facades of a motel, a gas station, and an airport, as well as the highly detailed homes, all made of the innovative material plastic, would snap together quickly to form a suburban sprawl, which could then be populated with *Plasticville* citizens designed to scale. Unlike other construction toys, which referred overtly to the built environment, *The Toy*, with its light materials, bright colors, and geometric forms, resembled a different kind of object—the kite.

The flying deployable toy, constructed, like the Eameses' product, out of colorful, lightweight materials, could also be found in stores in the early 1950s, rolled up in a long cardboard tube.[7] Kite

Box kite, used in the Gibson Girl survival kit during World War II (left); Kite, used as interior decoration in the Herman Miller Showroom, Los Angeles, Calif., 1949 (right)

flying had been a common recreational pastime for centuries but experienced a resurgence in popularity in the postwar era. Just prior to this time, the kite had proven valuable to the war effort. The *Gibson Girl Kite* was part of a rescue kit used during World War II. The kit contained a radio with a wire antenna that could be attached to the kite and flown above the rescue boat to emit a distress signal. Another kite, invented in 1943 by a curator of aviation at the Smithsonian Institution, also aided in the war. The image of an airplane was drawn on the underside of a large kite of the Eddy type, designed in the shape of a lozenge. The kite could be made to move in a controlled manner like a plane—banking right or left, climbing or nose-diving. It was used by the Air Force to train aircraft gunners to shoot at moving enemy targets.

Charles and Ray Eames, whose work always combined aspects of play with efficiency in design, embraced the kite as an object exemplifying these qualities. Their enthusiasm for kites dates back to the beginning of their practice in 1941 and is signaled by their inclusion of kites in the interior decoration of various projects, including the Herman Miller Showroom in Los Angeles, California of 1949. Kites were also incorporated in graphic design projects, such as the cover

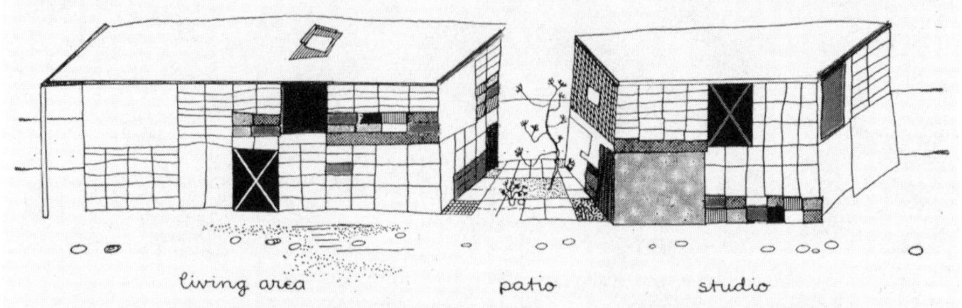

The Eames House, featured in "Life in a Chinese Kite," *Architectural Forum*, September 1950

of *Portfolio* magazine of July 1950, which shows a large pastel-colored flying toy. The designers were avid kite collectors, their collection comprised mainly of rare Asian kites brought back from the couple's numerous travels. Among the designs were Chinese kites representing fish, owls, butterflies, and dragons; Japanese kites depicting Kabuki faces, human forms, and painted landscapes; and multicolored Indian kites. The Eameses were consistently solicited to participate in kite-flying festivals and to submit material from their collection for publication.[8] By the end of the 1960s, they were loaning as many as one hundred kites to exhibitions across the U.S.[9] In 1978 the pair would produce *Kites*, a short film about the making and flying of a kite, as part of the longer film *Polavision*, which demonstrated Polaroid's instant movie camera.[10] *Kites* attested to the fact that even in the last year of their common practice, the Eameses thought the colorful toy a splendid object to behold.

"Life in a Chinese Kite," an article that appeared in *Architectural Forum* in September 1950, depicted Case Study House #8, the living and working spaces designed in 1949 by Charles and Ray Eames in Pacific Palisades, California. The home and studio would later become known as the Eames House.[11] The article described an architecture that made efficient use of light building materials yet presented a colorful and playful construction, just like a kite. The house was pictured as "light and airy as a suspension bridge—as skeletal as an airplane fuselage."[12] Although resting firmly on the ground, it was made of kitelike "standard industrial products assembled in a spacious wonderland."[13] The illustration accompanying the article featured the now-famous sketch of the house, drawn freehand, in which two boxes—accommodating the living room and the studio—stood side by side. With its thin, gridded skin enclosing airy volumes, the house resembled a Hargrave box kite, the first scientific kite structure, which consisted of two rectangular cells connected by straight, rigid rods.[14] Hargrave's was a stable kite whose lifting ability broke previous records and provided the basis for later aviation experiments.

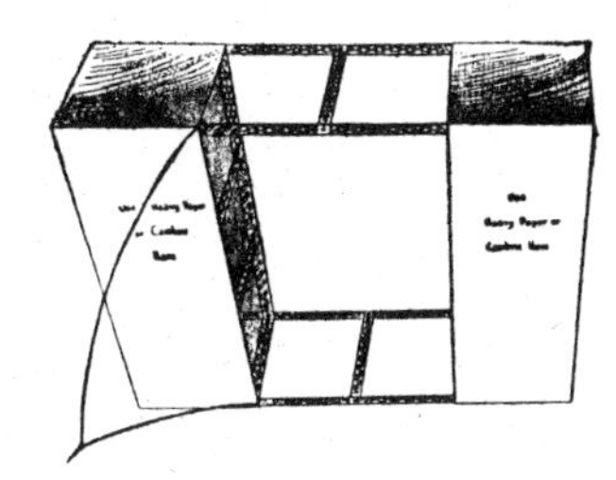

No. 252—Box Kite—Make frame construction with 12-in. girders and right angles, for binding together. Use heavy paper or cambric for planes. This is a practical model and can actually be used.

Manual of Instruction for the *Erector Set* (the American equivalent of the British *Meccano Set*), with directions for assembling a Hargrave box kite, 1915

Underscoring the playful nature of the Eameses' creative acts, the article explained that the "ready-fabricated parts" chosen from a catalogue of factory-made steel products were "bolted together like a Meccano set" to construct the house.[15] "Using this gigantic, clean-cut toy," the editors explained,

> Eames carved himself some 30,000 cu. ft. of space out of the air of the Pacific coastline.... Into the frame of his steel box kite he fitted sheets of glass, wire glass... asbestos, plywood or plaster in varying colors.... All combined to create an ever-changing play of light and shadow, a series of surprise vistas, of sudden planes of color suspended in mid-air.[16]

In the spaciousness, colorfulness, and playfulness of its architecture, the Eames House resembled a kite. But a material stronger than wooden dowels was needed to erect the structure. "How light is steel?" the *Architectural Forum* article asked. In response, Charles Eames recounted his effort to create a spacious steel-frame structure. He discovered the strength of steel: a 12-inch-deep truss could span more than 20 feet, a bent sheet could bridge 7 feet, and a 4-inch steel column could stand 17 feet tall. Turnbuckles and crossed wires held together the bolted frame. The architect's interest in aviation and marine equipment was reflected in his choice of materials. Referring to his use of "light steel"[17] in the house, he declared, "This is a material inspired by the daring of aviation engineers, rather than by the more timid techniques of traditional building."[18]

The spaciousness of the house, the seeming weightlessness of its building elements, and the modularity of its different industrial parts all related the home to the light structure of the kite. And like a kite, the Eames House was conceived as a kit of parts, which could be assembled and taken apart. The fixed set of factory-made elements that made up the home and studio could be combined and recombined to form alternative constructions.[19] In a similar manner, when constructing with the geometric panels of *The Toy*, different modular prisms could be assembled, arranged, and rearranged in endless variations, all of which presented open, colorful, flexible spaces. Thus, when playing with *The Toy*, one could practice building with a kit of parts, forming spacious enclosures with lightweight industrial elements, at a time when architects were experimenting with the modular parts of the factory-made house. Nevertheless, unlike other contemporary toys, *The Toy* did not present a domestic agenda; play with *The Toy*'s parts did not necessarily yield suburban homes—not even a small Eames House. Play was open-ended and involved combining and recombining parts, assembling interconnected webs and taking them apart. *The Toy* provided tools for experimentation. Its basic elements, when formed into prismatic building blocks—like structural particles—forced the player to tackle novel spatial configurations.

Buckminster Fuller, experimenting with the tetrahedron

FULLER'S MINIMAL UNIT

During the same years that the Eameses were developing *The Toy*, R. Buckminster Fuller (1895–1983) was playing with geometric elements as well, with the aim of developing the most efficient dwelling structure. While serving in the navy during World War I, Fuller had become aware of the processes used in shipbuilding and aircraft manufacturing. He witnessed how very large economic resources were poured into research, experimentation, and development of efficient wartime machines. It suggested to him the possibility of channeling similar scientific and economic resources into the design of the human environment in times other than war. During World War II, Fuller worked as technical advisor for the editors of *Fortune* magazine (1938–40) and served as head engineer on the Board of Economic Warfare (1943). From the vantage points provided by these posts, he observed the ineffective distribution of world resources. His observations and experiences during times of war led him to an unwavering involvement with researching light, transportable construction and its possible application for dwelling structures. Scientific research, he hoped, could be applied to the design of housing; by doing so, architecture could become a calculated endeavor, similar to the fabrication of a boat or an airplane.

From 1944 to 1946, Fuller found a supportive environment for his research at the Beech Aircraft Company in Wichita, Kansas. In a climate dominated by highly advanced technology and systems of production, Fuller and his team developed a circular metal housing structure that could enclose a large space very rapidly and that was so rigid it could withstand large environmental stresses such as tornadoes, hurricanes, and earthquakes. Dynamic forces—wind flow, heat loss, and changes in atmospheric pressure—were identified as the factors affecting the design of the house, encouraging the creation of a structure that, like an aircraft, could adjust to an ever-changing, unstable environment. Hovering above the ground to allow air to flow beneath the floor and equipped with a ventilator revolving on the rooftop, the shiny prototype appeared upon its completion in 1946 like an aerodynamic craft.

The Wichita House was an example of calculated research applied to dwelling. Although it aroused great interest among the professional community and the population at large, it remained a prototype.[20] Nevertheless, Fuller continued to pursue his dream of revolutionizing the building industry by attempting to design transportable dwellings that would adapt to different surroundings. Such a design, he believed, would provide a worldwide solution for housing.

After completing the Wichita House, Fuller published *Designing a New Industry*.[21] The "new industry" he described would entail a novel approach to architecture; one that would not replicate known forms but, through the handling of new, significant elements, would adjust to an unstable set of circumstances. The booklet contained a compilation of statements directed toward engineers who had come to see the prototype. Fuller sent to his friends the Eameses a copy of the publication with the personal dedication, "To Charles and Ray Eames, Number one colleagues in the building of this all important industry."[22] The building industry, according to Fuller, was lagging behind other trades in the amount of scientific and economic resources invested in its development. A renewed involvement with science, Fuller suggested, would inform and change design.

Man had always designed his housing using inefficient technologies and materials that were readily available, responding to emergency situations and survival strategies, maintained Fuller. As a result, houses resembled fortresses but could not perform as protective shelters. One of the critical features requiring change involved the weight of buildings. Weight was the main factor in the design of boats and planes yet did not come into play in the design of houses. "Architects are ignorant of the weight of their own buildings," he wrote, "though weight is the key to all industrialization."[23]

According to Fuller, failure to consider weight as a major factor left the house outside the forefront of industry. Housing should be thought of in terms of mobility and industrialization and should be valued on its "performance per pound."[24]

All in all, Fuller advised that the house should be lightweight, should be designed aerodynamically to resist environmental stresses and should be easily transported. In the attempt to find a structure that matched these criteria, the inventor-engineer developed his notion of "energetic-synergetic geometry." Working under the assumption that nature operated in the most efficient manner within a rationally determined coordinate system, Fuller devised a series of building blocks based on "nature's geometry."[25] Using such blocks in architecture, Fuller strove to achieve "maximal advantage in environmental control structures through effective energy accounting."[26]

Fuller looked for the minimal arrangement of vectors that would best represent the energy system of the universe. Packing small spheres around a central nucleus, he arrived at a fourteen-faced geometrical polyhedron that he called "vector equilibrium," as all its vectors were equal in length and equal also in the distance of each vertex to the center. This complex form was composed of octahedral and tetrahedral parts and, through a series of phases, could contract to become smaller polyhedrons, culminating in the octahedron, which in turn could be split into two tetrahedrons. The tetrahedron—a solid formed by four equal triangles—could not be split any further; Fuller labeled it "the minimum prime divisor of the omni-directional universe."[27] It was an indivisible unit and, therefore, a basic structural element. All other structures were a complex of tetrahedral transformations.[28]

According to Fuller, building with the tetrahedron as the elementary cell would lead to "maximum performance per pound of material invested."[29] He claimed that energy or force would always strive to push through the shortest distance—through the diagonals of a rectangle or a square. In light of this, triangular webs represented the most economical energy networks.[30] Thus, the engineer constructed systems or networks that derived from triangles. The projection of symmetrical, equilateral triangular systems composed of tetrahedra, octahedra, and icosahedra onto a sphere generated a structural system of great economy and provided maximum resistance to both external and internal forces.[31] This was the geodesic structure. By using networks of spherical triangles, Fuller formed the geodesic dome.

Fuller suggested experimenting with the tetrahedron and was keen on teaching tetrahedral principles, even to the very young. He wrote *Tetrascroll, Goldilocks, and the Three Bears*, a cosmic fairy tale for children (and adults) whose main protagonists were Goldy, the three bears, and a tetrahedron.[32] The book, published in 1975, was based on the bedtime stories he had told his daughter years earlier when she was a child. These stories revolved around a series of conversations in which the bears asked questions and Goldy provided answers. By relaying his thoughts to his young daughter in this engaging format, Fuller believed that "it would be possible to effectively induce that child's discovery of the most complex and profound phenomena."[33]

WACHSMANN'S BASIC STRUCTURAL COMPONENT

While Buckminster Fuller was developing his energetic-synergetic geometrical system, the German-born architect Konrad Wachsmann (1901–1980) was carrying out his own experiments with modular construction techniques.[34] In 1955 the Air Force commissioned him to develop a structural system for a very large airplane hangar. "The problem," Wachsmann later wrote, "was to develop a building system which, based on standardized elements, would permit every possible combination of construction, geometrical system, building type and span, expressed in a flexible anonymous design."[35] Furthermore, the hangar had to cover a very large area with no spatial impediments and be as light as possible, since the entire structure was to be made of collapsible parts that could be shipped in a compact (and thus economical) volume. Untrained professionals had to be able to assemble the structure, again very fast, into any desired arrangement.

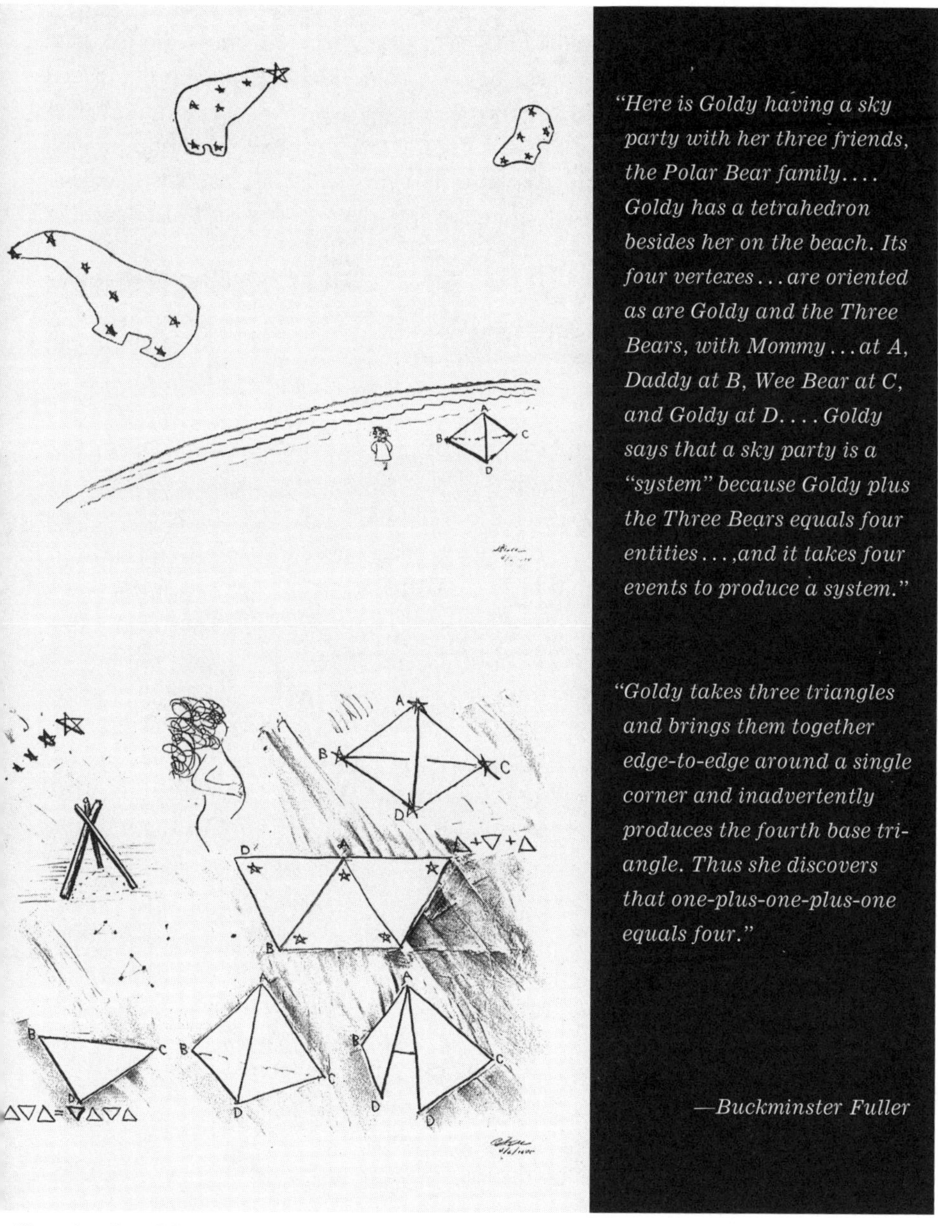

"Here is Goldy having a sky party with her three friends, the Polar Bear family. . . . Goldy has a tetrahedron besides her on the beach. Its four vertexes . . . are oriented as are Goldy and the Three Bears, with Mommy . . . at A, Daddy at B, Wee Bear at C, and Goldy at D. . . . Goldy says that a sky party is a "system" because Goldy plus the Three Bears equals four entities . . . ,and it takes four events to produce a system."

"Goldy takes three triangles and brings them together edge-to-edge around a single corner and inadvertently produces the fourth base triangle. Thus she discovers that one-plus-one-plus-one equals four."

—Buckminster Fuller

Illustrations from Fuller's cosmic fairy tale *Tetrascroll, Goldilocks, and the Three Bears,* 1975

Wachsmann, prototype for an aircraft hangar designed for the U.S. Air Force, 1955

The requirements for the airplane hangar led to the design of a space frame that, according to Wachsmann, necessitated the use of a basic cell—the tetrahedron. The choice of the tetrahedral cell was decisive; like crystals, each possessed rigidity along its three different axes, and like crystalline growth, the aggregation of many cells provided the larger truss with greater rigidity. The nodes of the tetrahedron used in the hangar's design were 10 feet apart, and a joint at the node was designed so that it could accommodate connections with up to twenty members. Based on this joint, different layers of the structure were built up to create a system capable of spanning approximately 500 feet, with 120 feet between supports. The space frame—a tetrahedral truss—could be shipped, assembled in the shape of a huge roof and substructure on columnar supports, and subsequently disassembled, quickly and easily.

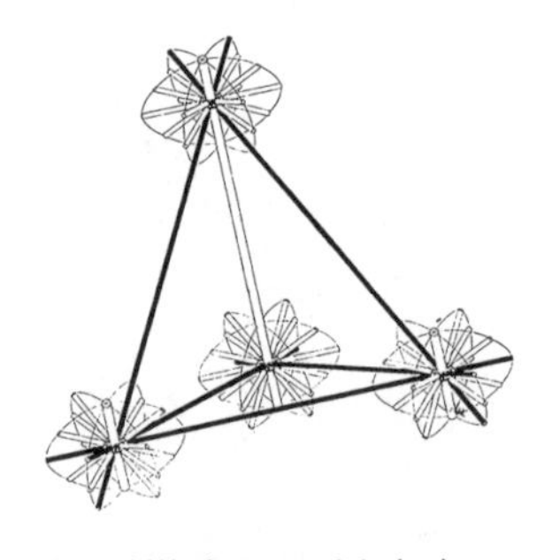

Konrad Wachsmann, tetrahedron—a three-dimensional structural principle developed for use in constructing an aircraft hangar, 1954

In his book *The Turning Point of Building*, Wachsmann wrote, "Building, which in the last analysis is a material struggle against the destructive forces of nature, obliges us to face the consequences of advances in science and the discoveries and inventions of technology, in order to identify, with every available aid and technique, the new laws of harmony between mass and space."[36] Wachsmann stated that "in discarding many of our old ideas about building, we have reached a turning point." "A new understanding of space," he maintained, "can only be achieved indirectly, by mastering materials, techniques and functions, by realizing creative decisions."[37] In this new project, he wrote,

> All preconceived opinion and notions of design are avoided and the best available resources and scientific knowledge freely applied. Quite indirectly, almost like a by-product, there emerged, at last, a structure capable of communicating a perfectly new spatial experience by technological means, while simultaneously expressing ideas of the conquest of mass and free dynamic spaces on a scale previously unknown.[38]

The turning point of building, then, implied building with light yet strong parts, allowing for modularity, transportability, and quick construction. All these characteristics were enabled with the use of a repetitive unit or cell. Manipulating such a module indicated a novel approach to design. Robert Le Ricolais, a French engineer and contemporary of Wachsmann's, commented, "Contrary to an ancient doctrine, in which the plan was divided into structural elements, the inverse procedure is being developed actually: the elementary cell is integrated, by proliferation in the construction."[39] Furthermore, despite their geometrical appearance, the triangulated space frames were immune to immediate mathematization; the calculations for an asymmetrical space-frame structure, for example, were inextricable to the extent that they could only be undertaken by specialists, he said.[40] Such structures in the early 1950s—an age preceding easy accessibility to the computer—could be best developed through models.[41] This process emphasized an architecture evolving from a tentative and additive procedure: a process of experimentation.

BELL'S LIGHTEST MODULE

Over half a century before Wachsmann developed his tetrahedral space frame, Alexander Graham Bell (1847–1922) had played with the tetrahedron. Indeed, the architect counted the construction of Bell's

Alexander Graham Bell, with tetrahedral cell structures, circa 1906

lightweight tetrahedral cell structures as an influential moment in structural design, along with the creation of monuments such as the Eiffel Tower (1889) and the Firth of Forth Bridge (1890).[42]

Bell, who invented the telephone in 1875, started to experiment with kites in the late 1890s. Initially undertaken for his own pleasure and amusement, his research grew into a serious affair when he realized it could have bearing on the contemporary "flying machine problem."[43] "The word 'kite' unfortunately is suggestive to most minds of a toy," Bell wrote, "just as the telephone at first was thought to be a toy."[44] Bell's proposal was "to construct an aerial vehicle large enough and strong enough to support a man and engine in the air, and yet light enough to be flown as a kite in a moderate breeze with the man and engine and all on board."[45] The idea of perfecting the kite came from the desire to reduce the velocity of the flying machine, so as to eliminate fatal accidents. A large, lightweight apparatus providing a small ratio of weight to surface area would be more subject to the forces of the wind and thus would be flown as a kite. In case of an accident in mid-air, the lighter machine would continue to glide with the wind until gently reaching the ground.[46] A heavier machine would be unaffected by the wind and thus would need to travel at a higher velocity in order to stay in the air, causing more accidents to occur, the thinking went.

In Bell's day, scientific kite flying primarily involved Hargrave's cellular box kite. Hargrave had proven that two rectangular cells separated by a large space but connected by a framework surpassed in stability all previous kites. By 1894 the U.S. Weather Bureau used his box kites to lift meteorological equipment to great heights. Bell began his own experiments with the Hargrave construction while attempting to create a kite of greater cell size. He discovered that the kite lost its lifting power as the larger rectangles became structurally weak and as the additional diagonal bracing needed to reinforce them made the kite heavier. During his test flights, Bell found that the horizontal surface of the kite resisted descent while the vertical one steadied the construction and prevented it from tipping over in the air. Bell concluded that diagonal planes might replace the horizontal and vertical ones of the rectangular box kite, creating a triangular kite of larger dimensions that could retain flying capabilities similar to the smaller, rectangular box kite. Such a kite had less lifting power, having only one horizontal plane, but had more strength and stability and was lighter, because one diagonal plane took the place of two used in the rectilinear model.

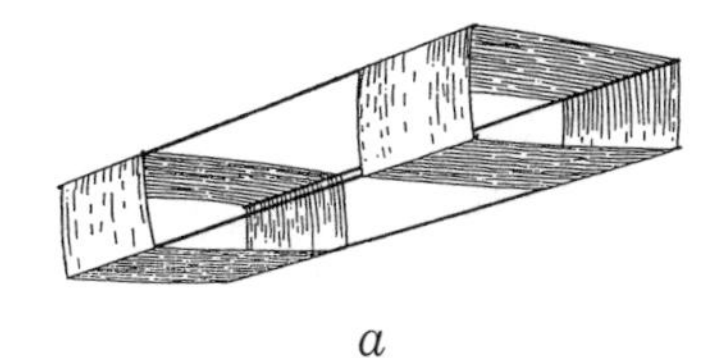

a

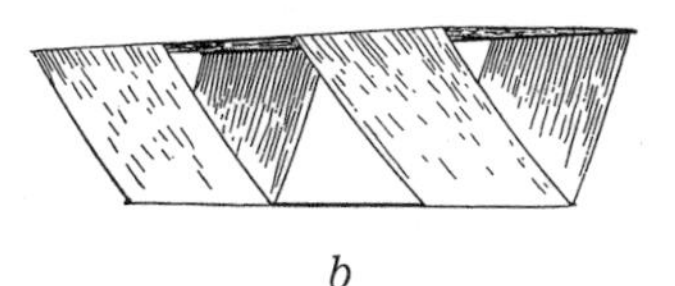

b

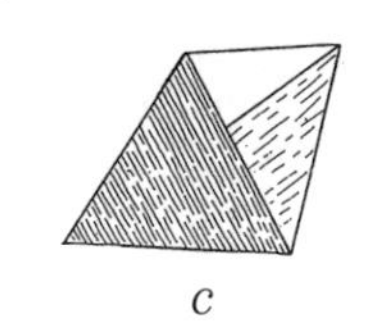

c

Diagrams of an a) box kite, b) triangular box kite, and c) tetrahedral cell kite

The sides of the triangular kite were initially formed by rectangular panes that required additional diagonal bracing when built at a large size. This again added to the weight of the kite and drastically reduced its ability to rise. In an attempt to improve the lifting ability of the kite, Bell changed the rectangular faces into stronger, triangular ones. Thus, six rods of equal length joined to form four adjoining equilateral triangles—the skeleton of a tetrahedron. The regular tetrahedron, he realized, modeled with equilateral triangles on every side, was a self-bracing form. This tetrahedral cell possessed extraordinary strength, even when made of lightweight elements, providing the maximum rigidity with the minimum amount of material. Bell wrote, "[The tetrahedron] is not simply braced in two directions in space like a triangle, but in three directions like a solid. If I may coin a word, it possesses 'three-dimensional' strength; not 'two-dimensional' strength like a triangle, or 'one-dimensional' strength

Bell, *White Flier*, a sixty-four–celled tetrahedral kite, after having flown at Baddeck, Nova Scotia, June 1903

like a rod. It is the skeleton of a solid, not of a surface or a line."[47] Covering two adjoining triangles with silk or another light material transformed the skeleton into a tetrahedral kite, or "winged cell." "The whole arrangement," the inventor maintained, "is strongly suggestive of a pair of birds' wings raised at an angle and connected together tip to tip by a cross-bar."[48]

Bell found that when a kite was enlarged while keeping its exact proportions, the ratio of weight to surface area increased; the heavier kite could not sustain its own weight, let alone carry an engine and a man.[49] He then speculated that a large kite could keep its lifting power if made by combining a series of small structures side by side. Bell formed a four-celled structure by connecting four tetrahedral unit cells at their corners, the entire structure taking the form of a tetrahedron. These four cells multiplied by four formed a sixteen-celled structure, and so forth. The ratio of weight to surface remained equal regardless of how many tetrahedral cells were combined together to form a compound kite. Furthermore, Bell found, "When these tetrahedral frames or cells are connected together by their corners they compose a structure of remarkable rigidity, even when made of light and fragile material, the whole structure possessing the same properties of strength and lightness inherent in the individual cells themselves."[50] After conducting numerous experiments, he observed that the equilibrium of the flying structure actually increased with the compound cell structure. A sudden gust of wind acting on one large surface could cause the structure to lose its balance, whereas a kite made of numerous cells would only sway slightly. He built kites of all forms based on the tetrahedral cell principle, and all had very

Bell, *Frost King*, built in 1905, here flown at Baddeck, May 1907

Bell (left) and his father, sitting inside the tetrahedral observation hut watching experiments with kites, Baddeck, August 1902. Here, Bell dictates his observations to a secretary.

good lifting power, were very stable, and when released returned very steadily to the ground.

Bell's kites were "enormous flying structures...really aerial vehicles rather than kites, for they were capable of lifting men and heavy weights into the air."[51] These kites could not be held by a single person by hand but were anchored to the ground with ropes wrapped around cleats, just like boats. He presumed there was a limit to the number of cells he could add before a detrimental effect would be observed, but it was a limit he never reached. In his laboratory, he put together 1,300 winged cells to form the 20-foot-wide *Frost King*, which in field tests conducted in 1905 carried its own weight plus that of several hundred feet of heavy rope and supported a man 40 feet in the air. The success of the *Frost King* convinced Bell that a large-scale tetrahedral construct was not impractical.[52] In December 1905, he proceeded to apply this knowledge of the tetrahedral cell to the design of a safe, stable flying machine.[53]

As strength and stability were desirable attributes for most earthbound, as well as airborne, structures, Bell recognized the broad applicability of the tetrahedral cell: "Just as we can build houses of all kinds out of bricks, so we can build structures of all sorts out of tetrahedral frames, and the structures can be so formed as to possess the same qualities of strength and lightness which are characteristic of the individual cells."[54] Using tetrahedral cells, Bell constructed three boats, windbreakers used to shield large kites on an open field, and an observation hut used during his kite flying trials. The largest structure he built with cells was a giant tower, located at the top of his experiment grounds at Beinn Bhreagh in Nova Scotia. The tower itself was a tetrahedron—its legs formed by a 72-foot equilateral triangle. Each leg was constructed of tetrahedral cells made of wrought-iron pipes joined with cast-iron connectors. The whole structure weighed less than five tons. Assembly of the structure was comparatively simple and was carried out in 1907 without the aid of cranes or scaffolding.[55] Bell's use of the tetrahedral truss some forty years before the advent of space frames attested to his realization of the advantages of cellular structure—in terms of rigidity, weight, and ease of assembly and disassembly, even by unskilled labor. In 1961 Wachsmann wrote of Bell:

Bell, tetrahedral cell tower at Beinn Bhreagh, Nova Scotia, August 1907

> Continuing his studies of the tetrahedral construction used in his air frames, in which weight was of decisive importance, he developed space-systems in the form of combinations of compression members, tie wires and stressed surfaces. All these were investigations which have become important in building research only recently.[56]

At the turn of the century, Bell had recognized the potential of the cellular structure and in his experiments with kites, blurred the distinction between invention and play.

PLAY

At that time all of us began to think
with our bare hands and even with blood all over
them, we knew vertical from horizontal, we never
smeared anything except to
find out how it lived.
Fathers of Dada! You carried shining erector sets
in your rough bony pockets, you were generous
and they were lovely as chewing gum or flowers!
Thank you!

—*Frank O'Hara, "Memorial Day 1950"*

In 1950, the Dutch historian Johan Huizinga's seminal book on play, *Homo Ludens*, was translated into English and published in the U.S.[57] According to Huizinga, *homo ludens* ("man the player") was essentially a fun-seeker, and play was a primary category of life, an integral part of culture, an addition to the natural course of events. Voluntary, never imposed by necessity, duty, or obligation, play embodied a sense of freedom. Consequently, it absorbed the player so fully and intensely that, inevitably, seriousness came into play. The notions of playfulness and seriousness were thus bound together in a fluctuating relation: "play turns into seriousness and seriousness into play."[58] And although seriousness was usually considered superior when contrasted with play, "play may rise to heights of beauty and sublimity that leave seriousness far beneath."[59] Huizinga defined the playground, or space where play took place, in the following terms:

> The arena, the card-table, the magic circle, the temple, the stage, the screen, the tennis court, the court of justice, etc., are all in form and function play-grounds, i.e. forbidden spots, isolated, hedged round, hallowed, within which special rules obtain. All are temporary world within the ordinary world, dedicated to the performance of an act apart.[60]

In play one was actually "stepping out of 'real' life."[61]

Playing with *The Toy*, combining one tetrahedron with another, adding, subtracting, measuring, and trying different configurations, circumvented usual strategies of design. It required making objects

Ray Eames and Konrad Wachsmann, wearing masks and dancing for the camera, possibly at the Eames studio, 1950

rather than planning them in drawn form. In a similar manner, the work of architects such as Wachsmann and Fuller entailed manipulating different materials over and over, working like a scientist in a laboratory until a successful experiment was achieved. Bell's long-term research on cellular structures took this form as well; for years he would go outdoors with his assistants and fly kites. His was not a theoretical enterprise, and only through experimentation did he determine the supremacy, in strength and stability, of the tetrahedral cell structures.

The Eameses also treated their practice as a laboratory and their studio as the place for experimentation. Examining numerous options for every design, closely comparing performance and form, they used their workshop as a testing ground for furniture, film, architecture, and toys. As educators, they wished to extend to others their design approach, and *The Toy* can thus be seen as providing the elements necessary to lead a successful experiment, anywhere. This search would not yield a determinate building structure, such as a house or a school; instead, the combination of parts might lead through playful endeavor to unexpected and innovative results.

The ingredients of this experimentation were light, transportable, modular, and easily assembled, which when combined formed rigid, airy, inhabitable structures. Like molecular particles, the tetrahedral cells bonded to one another to make a larger, stronger system. Other architects started experimenting with similar constructs, using triangular nets in the design of buildings. Louis Kahn combined the tetrahedron, which he called a "hollow stone," vertically, in his project with Anne Tyng, Tomorrow's Town Hall (1952–58). Among the numerous projects with triangular networks are Buckminster Fuller's Octet Truss, presented at the MoMA in 1958, and the Japanese architect Kenzo Tange's space-frame projects of the 1960s.[62] Toying with tetrahedrons presented open-ended play.

On the pages that preceded "Building Toy" in the July 16, 1951 issue of *Life* magazine, a large color photograph showed a huge orange ball of fire exploding amidst a pink sky. The short article "Atomic Bomb over Nevada" described in matter-of-fact terms the workings of the A-bomb, as demonstrated in a recent test explosion. It explained how, following a detonation, a ball of fire begins to rise:

> [In] a few seconds [it] will be climbing at a speed of almost 200 mph. The reddish column beneath the fireball is formed by dust sucked up from the ground below. This dust churning upward with the fireball will later help form the familiar mushroom cloud of atomic explosions. . . . The bomb's fission products (fragments of plutonium or uranium atoms) are already so intensely radioactive they can cause air to glow. And since they were vaporized in the first instant of the explosion, they may have been carried even higher than the fireball itself to produce the mysterious purple haze at the top.[63]

On the pages that followed, the orange and pink panels of *The Toy's* light, prismatic forms cast their glow. The shift from seriousness to play, from destruction to construction, is abrupt but revealing. Could handling *The Toy* imply that one could reclaim the sky as a playground for a different kind of experiment?

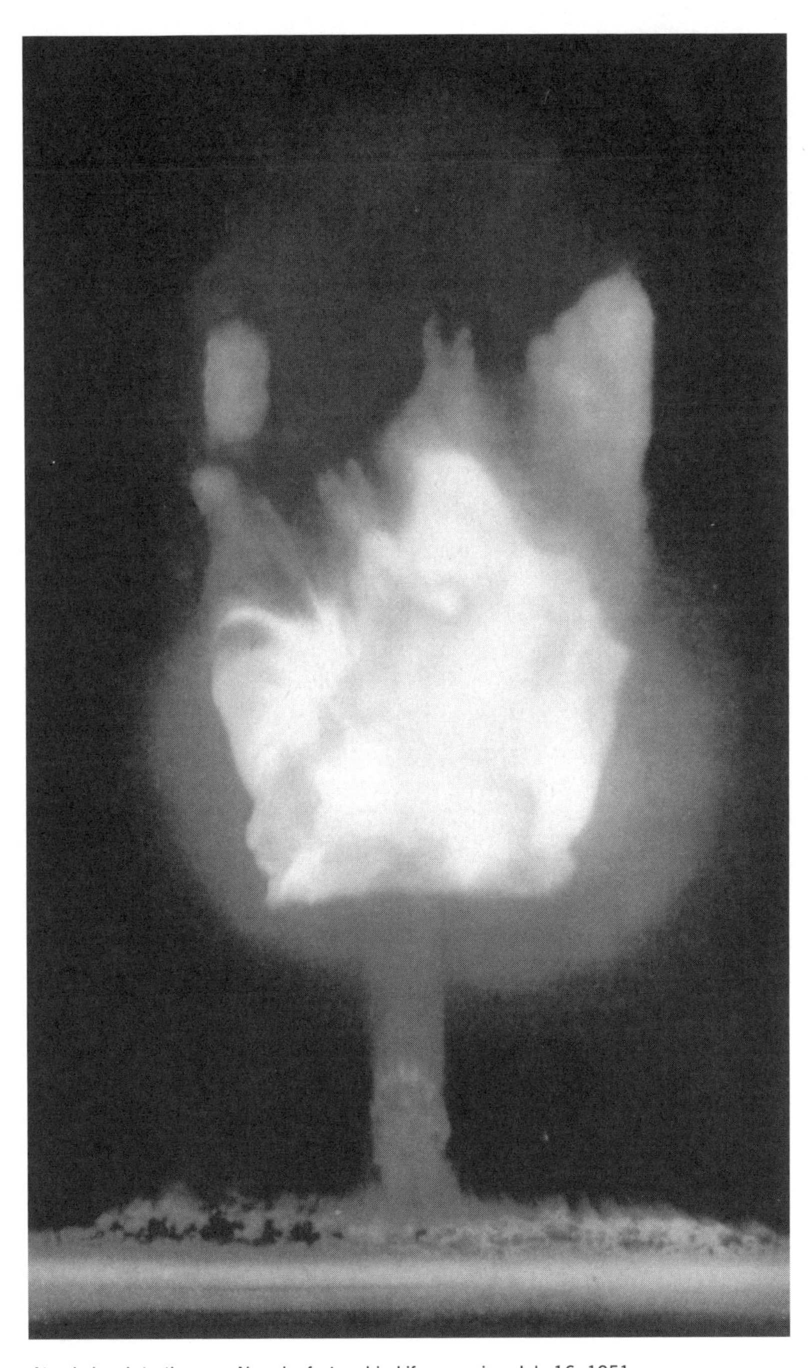

Atomic bomb testing over Nevada, featured in Life magazine, July 16, 1951

MISSION 66

JEANNIE KIM

In 1958 President Dwight D. Eisenhower established the Outdoor Recreation Resources Review Commission (ORRRC) as an adjunct to the Department of the Interior. The commission was charged with gathering information on the "great outdoors" and producing a national policy on recreation. Published in 1962, the findings of the ORRRC concluded that, no longer just a fad, outdoor recreation had become "a component of the national character."[1]

Outdoor recreation found its way into the "national character" in the preceding decade, largely as a result of the efforts of the National Park Service. In 1955 the park service launched an effort to modernize the park system and update its facilities in response to the anticipated influx of visitors to coincide with the celebration of its golden anniversary, in 1966. This effort, organized under the banner of "Mission 66," would allocate over $670 million to build and repair roads, trails, landscapes, and facilities within existing parks, acquire new land for future parks, and expand the number and scope of duties of the park service staff. In revamping its facilities, the National Park Service would also alter the way America thought about the role of man-made intervention in the natural environment.

SITE

With a substantial donation of tent villages on behalf of the Rockefeller family, the National Park Service (NPS) was established in 1916 and charged with safeguarding America's natural wonders. Through the parks movement of the mid-nineteenth century,

Automobile congestion at the entrance to a national park, 1956

which typically embraced horticultural reform as a way to improve urban life, nature tourism had become an established institution.[2] As natural oases emerged within the gridded landscape of American cities, recreation became a new form of leisure.[3] Following World War I, recreation moved out of the cities as automobile owners embarked on weekend and holiday excursions into the American countryside.[4] The modern highway encouraged mobility and came to be associated with the very concepts of democracy and freedom. The highway system contributed to an unprecedented growth in attendance to the NPS's parks, campgrounds, and natural wonders.

The marketing of the national public landscape toward the private car quietly transformed what was described as a democratizing gesture into a form of leisure that was directed toward a specifically white, middle-class nuclear family.[5] Car and family became necessary equipment for outdoor recreation, and the ideology of the park system—as demonstrated in the rhetoric of official NPS publications and tourist brochures—promoted the notion of leisure as a cultural curative for the ills of democratic society, a sort of booster shot for Americanism.

Faced with the daunting prospect of accommodating 80 million automobile vacationers within the already overcrowded national park system by 1966, Director Conrad Wirth assembled a team of researchers. Known as "the Squad," this team—composed of many who had served with Wirth in the Civilian Conservation Corps (CCC)—set upon an "objective and scientific study" of the needs of the NPS. Mission 66, as the study was called, was immediately

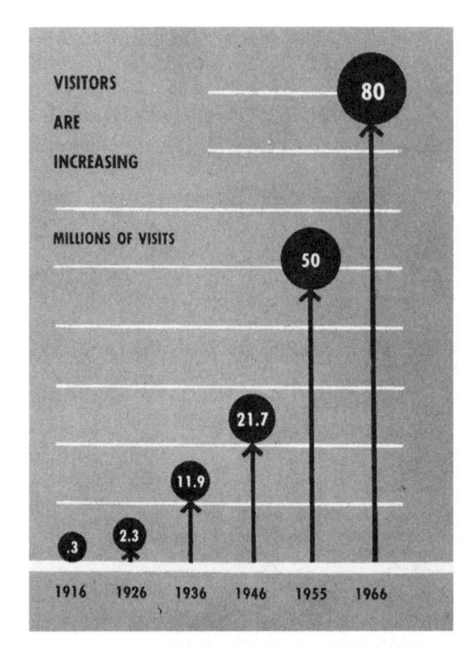

National Park Service, projected visitorship to national parks, 1956 (left); National Park Service Director Conrad Wirth (second from left), Vice President of the American Automobile Association Russell Singer (right), and other guests, celebrating the kick-off of Mission 66 with a meal of bison meat, 1956 (right)

approved by President Eisenhower. The kick-off dinner for the research effort, held in February 1954, was cosponsored by the American Automobile Association and attended by Wirth and Russell Singer, Vice President of the American Automobile Association (AAA). The relationship between the NPS and the AAA was both necessary and monetary, as the parks were increasingly marketed toward families with automobiles and the profits of the AAA increasingly depended upon driving vacationers.[6]

The Squad began its research with the directive from Wirth to study the legislation that dealt with the NPS and quickly found that no new provisions had been made since the service's annexation from the CCC in 1936. The Squad then produced reports on the way park services were handled in England and Switzerland. Favoring the English approach, the team wrote about these precedents in terms oddly nostalgic for the period in the nineteenth century when outdoor recreation was associated with health spas and convalescence.[7] Once one walked long enough and far enough in the national parks, Wirth waxed poetically in the preface to the report, "there would be no trouble which at the end of the day would not look different and feel lighter."[8]

Wirth's expressions of concern for the mental and physical well-being of the park visitor were brief and uncharacteristic. Much of the rest of the findings of the Squad highlighted the fact that the National Park Service stood to make significant financial gains from a mass influx of postwar visitors. Well-intended concern for the health of visitor was soon overshadowed by the knowledge that the NPS made $116 million ($29 million in federal taxes alone) at Grand Canyon National Park in 1954.

Following a report on the general gains of Mission 66 for the NPS system, the Squad embarked upon a series of case studies of various parks. These reports took the form of amateur ethnographies, followed by "creative suggestions" for improving the accessibility and services within the park were largely dismissed by park administrators. The findings of the first pilot study, on Mount Rainier Park, demonstrate the struggle of the Squad—and of the NPS generally—to find a balance for Mission 66 that fell somewhere between the preservation of the natural landscape and its efficient use. Suggestions included the following:

> Women want good trails, trails that they can walk on in high heels. Trails to points of interest should be hard surfaced for all-weather use and smooth enough for high heels....
> Mechanical and audio-visual devices cannot answer visitor questions. Gadgets cannot replace people....
> There should be more life-size exhibits in historical parks. Museums are only aids to the "thing" itself. People are showing an ever-increasing preference for life-size reconstructions in the places they visit....
> A visitor service program should provide simple information. Give opportunity for an experience in which each person directs himself....Do not provide for artificial types of recreation—that is, recreation that requires man-made facilities to engage in it.[9]

After presenting their findings at the Regional Directors Conference in April 1955, Wirth hired more researchers, and more pilot studies were undertaken. The Chaco Canyon team provided a test case for the future reception of Mission 66 with their study of the new park in New Mexico. Superintendent of the as-yet-unopened park Glen T. Bean had been planning a road down to the canyon floor. Based on the Mount Rainier findings that favored self-discovery and the cultivation of a "frontier spirit," the Squad counter-proposed bringing the road only a short distance away from the highway to the rim of the canyon—a move that would draw in chance visitors—and preserving the frontier experience by constructing a natural foot trail into the canyon. Using the Squad's own arguments against them, Bean cited the imperative to create hard-surfaced trails suitable for "women in heels," and the master plan remained unchanged. Frustrated by Bean's dismissal, the Chaco Canyon team abandoned their study.[10] Despite their hopes for a welcome reception at Chaco Canyon because of the novelty of the park, the superintendent's dismissal reflected the general attitude of the park establishment toward Mission 66.

In November 1955, Wirth hired the public relations firm of Audience Research in Princeton, New Jersey, to survey Americans about their awareness of the NPS. Their findings furthered Wirth's belief that the parks were not serving their public. Audience Research found that 90 percent of park visitors arrived via personal car; it also found that 64 percent had at least one child and that 95 percent were white. Of more importance to Wirth and the Squad were statistics indicating that the public could name only two national parks, that 34 percent of visitors were on route when they accidentally came upon the parks they visited, that 84 percent of visitors had at least a high school diploma, and that more than half of visitors came from urban areas.[11] After receiving the report, Wirth turned it into media-friendly epithets and set about publicizing it, garnering attention from *The New York Times* and *Saturday Evening Post*, and eventually earning the support of President Lyndon Johnson, who expressed his conviction that Mission 66 had to "secure the beauty of America."[12]

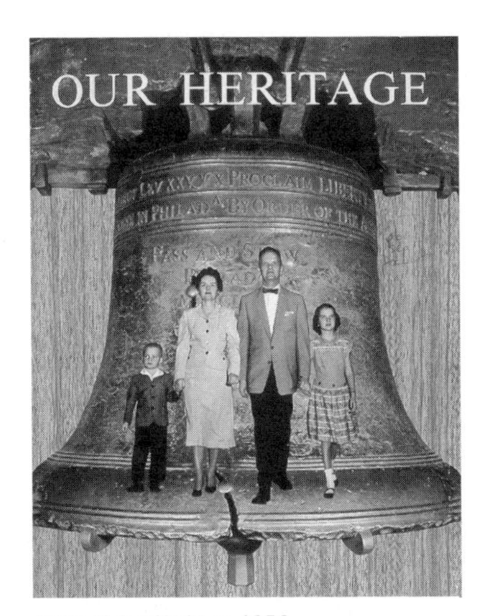

Cover of *Our Heritage*, 1956

Wirth published the goals of Mission 66 in *Our Heritage*, a publication dedicated "to the liberty of the individual" that was underwritten by Standard Oil and the AAA.[13] The twenty-four-page pamphlet produced by the Department of Forest Economics in 1956 sought to address "what the National Park System means to Mr. and Mrs. America and their children."[14] From its opening image, featuring the all-American family floating in front of the Liberty Bell, the pamphlet made the National Park Service's mission clear: the efforts of Mission 66 would preserve the role of the park system as a producer of national character and good citizenship. The NPS continued to maintain Wirth's nineteenth-century vision of the park system as a respite from the ills of urban life, writing in another government publication of the same year, *The Need for Mission 66*, "Here people cast off the cares, escape the sounds, and get away from the grind, tensions, and monotony of routine living. Parks are a spiritual necessity, an antidote to the high pressures of modern life, places to regain spiritual balance and find strength."[15] But as the introduction to *Our Heritage* made clear, parks were also presented as places that were "owned by all Americans," where "one [could] gain a new sense of American valor and courage."[16] The park service explicitly promoted a view of the parks as federally-funded realizations of the democratic vision—a public space that embodied the tenets of a government founded by and for the people:

> Where else do so many millions of Americans, under such satisfying circumstances, come face to face with their Government? Where else but on historic ground can Americans better renew the idealism that prompted the

> patriots to their deeds of diplomacy and valor? . . . The park visitor is a better man afterward, a more effective and a more productive member of his society.[17]

This passage couches veneration of the national park system as part of an allegiance to the democratic state. An emerging American secularization, earlier manifest in the view of the natural landscape as a spiritual site, was being countered, in part, by a newly configured nationalistic theology, represented on the pages of *Our Heritage* and other publications.

Updating the NPS's original role as the sentinel of our natural wonders, *Our Heritage* promoted the idea that the parks—and use of the parks—were good for government. In a climate of postwar consumerism, visiting the parks was discussed as yielding, not material products, but "the great cultural and inspirational products of knowledge, refreshment, and aesthetic enjoyment equally needed by all people."[18] In the pages of *Our Heritage*, the goals of Mission 66 were reduced to eight objectives:

> 1. Additional accommodations and services adapted to modern recreational needs.
> 2. Protection of park resources.
> 3. Services to make the parks more meaningful and to encourage protection of the parks through visitor cooperation.
> 4. Field staff to manage areas and maintain facilities.
> 5. Adequate living quarters for field employees.
> 6. Extinguish grazing rights and competing water rights.
> 7. Coordinate the national recreation plan between Federal, State, and local interests.
> 8. Protect wilderness areas and encourage education to prevent them from being impaired.

INTERPRETATION

Based on the findings of the Squad, *Our Heritage* concluded that Mr. and Mrs. America were becoming increasingly urbane and that the NPS needed to respond accordingly. The impact of a projected 80 million visitors in 1966 would create obvious traffic and accommodation problems, but its authors also speculated,

Suggested combination of live explanation and audio-visual device, 1956

> These problems [would] multiply when travel include[d] more city-bred people unfamiliar with wilderness ways...; more people whose rising level of education demand[ed] more knowledge and guidance; and more citizens accustomed through the press, radio, television and motion pictures to the professional and graphic presentation of knowledge.[19]

The image of sophisticated city folk with high intellectual demands ran counter to the image of Teddy Roosevelt's nineteenth-century Rough Riders making their way through the American wilderness. Faced with what was essentially a graphic design challenge, the federal government granted the NPS permission to embark on a series of ambitious studies to address interpretation within the park system.

A rivalry borne out of redundancy between the NPS and the United States Forestry Service (USFS), both housed within the Department of the Interior, had produced varying standards of presenting information throughout the nation's system of parks.[20] From rustic burnt-wood signposts to reflective highway signage, the graphics of the NPS encouraged accusations by the USFS that it waffled between entertainment and pseudo-scientific research in the presentation of knowledge—its most valuable product. With

The first National Park Service visitor center, designed by Cecil Doty and built at Carlsbad Caverns National Park, N.Mex., in 1956

funds set aside for Mission 66, the NPS established a think tank in Washington, D.C., that also served as an audio-visual laboratory to test the latest orientation devices to be used in the parks. In addition to developing new methods of delivering information, the Division of Interpretation would determine content, promising a "more scientific approach" than had been taken in years past. The NPS had previously relied upon volunteers and interested research institutions for its information, which resulted in a rather uneven quality of scholarship. With the formal founding of the think tank, research would now be systematized and, perhaps more importantly, controlled through grants, fellowships, and a program for scholars-in-residence.

Questions regarding what constituted appropriate orientation devices had plagued the NPS from its beginnings. In 1929, a series of guided automobile caravans led in procession by an interpreter speaking from his rooftop via a bullhorn wound their way through

Mesa Verde and Yosemite.[21] Electrified maps, slides, and other audio-visual media were introduced throughout the major parks by the 1940s. Through the efforts of the Division of Interpretation, visitor-activated audio-visual devices including hand phones and slide movies came into wide use by the 1950s. Following studies of retinal activity and attention span, the agency recommended limiting movies to thirty slides and ten minutes of narration to promote optimal knowledge absorption. Further stylistic guidelines suggested that the programs should begin with a reference to the NPS and include slides of a man in uniform helping visitors enjoy the park.[22]

In 1967, while surveying his accomplishments upon his departure from the National Park Service, Wirth launched a study of its interpretive activities. The study was so scathingly dismissive of Mission 66 efforts that he commanded that all of the reports be destroyed after regional park directors read them over a single weekend.[23] Like the National Park Service's in-park presentations, its publications typically fell somewhere in between entertainment and education. Despite the efforts of the Division of Interpretation, the ideological message of NPS would remain unclear. Unable to precisely define its audience, the division became a testament to the perceived failure of Mission 66.

THE VISITOR CENTER

The National Park Service's interpretive efforts included the refinement of a new building typology: the visitor center. Created as an effort to reinterpret the role of the museum within the park, the visitor center was proposed as a new building solution that could accommodate a range of programmatic demands while allowing for a separation between park interpretation and entry. Previously, the New Deal had funded museum construction within the parks, and by 1939 there were seventy-six museums in the park system. Through the efforts of Mission 66, parks were forced to address the problem of interpretation and the role of the building entered by park visitors before the discovery and self-exploration promised by the park itself. Most importantly, the visitor center would provide orientation for otherwise hapless visitors:

> The visitor center is the hub of the park interpretive program. Here trained personnel help the visitor start his trip and with the aid of museum exhibits, dioramas, relief models, recorded slide talks, and other graphic devices, help visitors understand the meaning of the park and its features, and how best to protect, use, and appreciate them.... [P]arks lack visitor centers today, and a substantial portion of park visitors, lacking these services, drive aimlessly about the parks without adequate benefit and enjoyment from their trips.[24]

The first visitor center was at Carlsbad Caverns National Park in New Mexico. The relatively modernist building, designed in 1954 by the NPS's in-house architect Cecil Doty, demonstrated an embrace of "the open plan of the future" through the removal of interior partitions and standardized, yet operable, fenestration.

With flat roofs, free plans, concrete construction, interior courtyards, and contextual embellishments (indigenous wood finishes, live dioramas, etc.), the style came to be known as "park service modern" and was regarded as "a reinterpretation of the long-standing commitment to harmonize architecture with park landscapes."[25]

Modernism in the form of asphalt, precast concrete, and aluminum windows was accepted by the NPS in the name of efficiency, not ideology. Like the use of scientific analysis to predict the desires of park visitors, the role of modernism in the park system was to efficiently orient the park visitor. In 1957, occasional NPS architect Emerson Goble expressed his dissatisfaction with the (pre-Mission 66) state of NPS architecture:

> Let us not decide... great and sympathetic architecture cannot exist. I shall have to insist that the effort to achieve or acquire great architecture has almost never been tried. The whole habit of thinking in the parks is the other way. We have not dared to let man design in the parks; we have not asked to see what he might do. We have slapped his hand and told him not to try anything.[26]

For Goble, and for others, Mission 66 represented an opportunity: it challenged the NPS to envision modernism as an indispensable part of its identity.

The new building typology of the visitor center was announced in Mission 66 literature, celebrated in NPS propaganda, and quickly adapted to fit park service ideology. Whereas previously the task of orientation was left to small museum-like buildings near entrance gates, visitor centers heralded the arrival of something different. Ambitious modern structures that housed information and interpretative facilities, exhibits, rest areas, and administrative offices,[27] they included souvenir stands, cafeterias, audio displays, and even small auditoriums in the larger parks.

Flat roof, featured in the National Park Service's pamphlet *Mission 66 in Action*, 1957

The relatively undefined program of the visitor center focused around the term "center," reflecting the desire of NPS planners to centralize interpretation within the parks. The reasoning behind this mode of centralization was related to planning ideas found in other new typologies, including shopping malls, corporate campuses, and industrial parks.[28] Previously, "park village" planning had been more decentralized, disseminating facilities within the park in order to make them as inconspicuous as possible. It is not by coincidence that it is precisely after the war that the planners at the NPS began to desire a central "control point" to promote a more efficient pattern of public use and straightforward "orientation." In this ambiguous manner, the visitor center sought to give form to the task of interpretation within the parks with a new typology aimed specifically at "orientation."

ROAD

The visitor center was meant to act as a step-down converter between the pace of urban life and the experience of nature without obviating the need to see the park at all. NPS historians favored siting visitor centers "right on top of the resource" so that visitors could see everything from the visitor center. This became particularly important at sites without physical features, but during the postwar era, also became a way of drawing unsuspecting cars in from the highway. The visitor center became "a signpost to attract the aimless visitor within." Tellingly, the planning of the visitor center typically began with circulation and flow diagrams.[29]

As part of Mission 66, the National Park Service committed itself to the completion and renovation of the national parkways by 1966. The Blue Ridge, Foothills, George Washington Memorial, Natchez Trace, and Colonial parkways, and the subsequently abandoned Chesapeake and Ohio Canal parkways were described as "elongated parks bearing a road as a major feature," and therefore they had to be maintained as natural attractions.[30] The roads would be developed with the aid of the 1954 Federal Highway Act but with the added NPS caveat that they would be designed to "satisfy the curiosity of the traveler."[31] The precedent for the parkway as an environment of instruction came from the Blue Ridge Parkway, completed during the New Deal period by the CCC, under the direction of the NPS.[32] Armed with this experience, the NPS set out to transform the notion of recreational driving for the middle class at mid-century.

Originally intended to link all of the parks along the length of the eastern seaboard, the Blue Ridge Parkway was scaled back to join just two—the Shenandoah and Great Smokey Mountain National parks, located in Virginia and North Carolina, respectively. This first national parkway became a 470-mile exercise in land control and graphic design. The parkway was designed to present nature to the automobile passenger in a way that enhanced the experience of driving, allowing the privacy of the nuclear family riding in a single car to regain its primacy as part of the equation of park visitorship. The

Lincoln Memorial, Gettysburg, Pennsylvania, 1962

NPS secured control over the land surrounding the parkway and set design guidelines proscribing the use of rustic split-rail fences and regarding the pasturing of cows (among other things) by private citizens whose land happened to border the roadway. The road itself was designed with curves that intentionally limited speeds to 35 or 40 miles per hour and provided drivers and passengers with "long looks" and verges—places where the driver's eye comes to rest after looking up from the road.[33] Despite its random rustic appearance, the parkway was an exercise in repetition. Signage was standardized along the parkway and a rhythm of overlooks, short trails, small museums, and local parks established to provide a diverse visual and recreational experience calibrated in 30-mile increments.[34]

During the early 1960s, shows were performed on the edge of the Blue Ridge Parkway with men in crooked hats and women in flowered dresses evoking a fictionalized "hillbilly" culture that appeared to persist uninterrupted by the insertion of the sinuous roadway. By the time Mission 66 took over the renovation and modernization of the parkway in 1958, the "spontaneous" performances had disappeared. The NPS replaced living history with local craft

Auditorium with operable window-walls and panoramic view of the Gettysburg battlefield

museums and interpretative centers. By the fiftieth anniversary of the NPS in 1966, Mission 66 was responsible for either building or renovating more than 2,000 miles of road within the national parks.

More than any other Mission 66 accomplishment, the apparent commitment to the creation and maintenance of these linear parks invited the derision of environmentalists and criticism of the program as one that was dedicated to the automobile. Similarly, the apparently welcome sponsorship of the AAA and Standard Oil contributed to the reputation of Mission 66 as a "road and big development program" that had done little for the resident plants and animals and "nothing at all for the ecological maintenance" of the park system.[35] The introduction of modernist visitor centers into the park system furthered the cry that Mission 66 had alienated nature within the national parks.

THE LINCOLN MEMORIAL

As part of the National Park Service's campaign, the planners of Mission 66 hired a group of modernist architects, including Richard Neutra, Eero Saarinen, Anshen and Allen, and Mitchell-Giurgola, to design visitor centers for various parks beginning in 1956 with Anshen and Allen's Quarry Visitor Center at Dinosaur National Monument in Jensen, Utah. This brief embrace of a modernism that went beyond the use of aluminum windows and precast concrete, was anticipated in the park service's 1957 brochure *Mission 66 in Action*, in its depiction of a streamlined modern visitor center with viewing terrace and flat roof. Although Wirth's preference for modernist architecture was not explicitly stated, his tenure as director of the NPS produced works that departed from the rustic style created previously and were praised for their efficiency, embrace of technology (particularly high-speed elevators), and laconic presence.[36]

For Wirth, modernism represented a form of contextualism that would allow the visitor center to blend into the landscape through a lack of embellishment rather than an imitation of nature. In fact, that rustic architecture drew too much attention to a building intended to be purely functional was a defense used by Wirth to reject Frank Lloyd Wright's proposal for a restaurant at Yosemite of 1954.[37] In September 1955 Wirth distributed a recommendation to NPS field officials that suggested his aesthetic disposition:

> [P]ark structures are to conform, to some extent, with the trend toward contemporary design and the use of materials and equipment accepted as standard by the building industry. However, restraint must be exercised in the design so that the structures will not be out of character with the area and so that the structures will be subordinated to their surroundings.[38]

Richard Neutra shared Wirth's belief that modern architecture could blend into the landscape. In 1958 he was commissioned to design a visitor center on the site of the Battle of Gettysburg. An Austrian immigrant, Neutra himself was surprised by the commission, but his association with planner Robert Alexander and his previous work on government commissions (such as the Federal

Housing Administration Chavez Ravine Housing complex of 1949) apparently convinced the NPS of his civic commitment.[39] The program of the Gettysburg visitor center was largely determined by the presence of several cyclorama paintings completed in the 1880s by the French painter Paul Dominique Philippoteaux. After the initial development of the possible site strategy and massing by Doty, the plan and project were handed over to Neutra and Alexander.

A controversy over the interpretive and landscape strategies of the building quickly ensued within the NPS. Living history had become a temptation at Gettysburg, a significant historical site with no striking physical attractions.[40] Academic historians objected to the use of a site for ends only indirectly connected to the historical events that took place there. Debate arose over what costumes people would wear and what the general store would sell at the site.[41] At Gettysburg, the questions regarding the programming of the park rekindled the earlier NPS battle over preservation vs. use. Some in the park service (typically the conservative veterans of the CCC) advocated restoring the battlefield at Gettysburg to the exact conditions at the time of battle but resisted the introduction of any living history exhibits. Gettysburg was a battlefield and not a park, they argued; cannons should be moved out of their ceremonial positions and back to where they were during battle. Monuments that had been erected on the battlefield should be removed, and lawns should be left unmowed. "We're not in the lawn business," a local NPS historian maintained, "We're in the business of preserving a battlefield. With lawns you get a problem; they become recreational areas for the town, and this creates the wrong atmosphere.... The history should be enough, recreation profanes it."[42]

This debate between preservation and use was not a new one within the NPS. Whereas the advocates of use argued that parks should accommodate the changing needs of an increasing visitorship, those in favor of preservation viewed the maintenance and, more often than not, recreation, of "historic" conditions as the greatest service to the future of the parks. In the end, the preservationists won out—to the detriment of park visitorship and the peril of the Neutra building. The cannons were put in their proper place,

"Park rustic" fences and landscaping along the interior ramp of Neutra's Lincoln Memorial Visitors Center (left); Indigenous plants and window-walls elide the boundary between interior and exterior (right)

monuments removed, grass allowed to grow, tennis courts removed, and kite flying prohibited. Curiously, the conservative attitude that guided the reworking of the site itself did not extend to Neutra's stripped-down modernist visitor center. The project survived and was finally dedicated on November 19, 1962.[43]

The building itself is a photogenic example of Neutra's commitment to the belief that stripped-down, functional modernism could fade into the landscape. Sited parallel to and somewhat sheltered from the street, the building was further camouflaged by an irregular masonry wall built of local stone. Mixing the vernacular with white-wall modernism, landscape with interior, the contextualism that characterized Neutra's late work seems oddly in keeping with earlier aesthetic requirements of the "rustic" within the NPS. With the exception of the (remotely located) parking lot and the pedestrian ramps that link interior and exterior, the Gettysburg Visitor Center is marked by Neutra's familiar residential vocabulary of mirrored pools, operable window-walls, rough stone, glass corners, and exposed beams. Driven by the necessary linearity of the ramp, the visitor center is a physical realization of the National Park Service's interpretive sequence, drawing the visitor in from the parking lot, past a map of the site, through the cyclorama, into a room with a panoramic view of the battlefield (where a narration of Lincoln's Gettysburg Address plays), and then down a pedestrian road to the battlefield itself. Orientation is rendered inevitable by the control of circulation, and the visitor arrives at the park's final destination only after being properly programmed with the appropriate information.[44]

The building of national character after World War II enabled new cultural forms that reflected the complicated liberalism of the postwar era. In his account of the cultural history of abstract expressionism—a period of modernist art that roughly coincided with the tenure of Mission 66—Serge Guilbaut adeptly recognizes Arthur Schlesinger, Jr.'s 1949 book *The Vital Center*[45] as a harbinger of the new liberalism of the Cold War, which positioned itself between the ideologies of the conservative right and liberal left while leaving room for the avant-garde in the center. The result of locating the avant-garde between the right and the left, Guilbaut notes, was a notion of individual freedom "in a heavy cloak of ideology."[46] Avant-garde artists became politically neutral "individualists" as their work was assimilated and utilized by politicians in the name of "liberal" ideology.[47] A similar ideology was at work within the national park system, where individual experience was subsumed by a greater democratic collectivism. This complex combination of control and freedom led to the vague position of the NPS when addressing questions of orientation, centralized planning, and federal policy.

Unlike abstract expressionism, the modernism briefly adopted by the NPS was inherited through the legacy of the International Style and therefore did not have the same associations with radical politics forged out of alienation. The desire to set themselves against the totalitarian uniformity represented by the Soviet threat was intrinsic to both, but the notion of freedom—the specific individualism advocated by the NPS and unwillingly subscribed to by the artistic avant-garde—was necessarily mediated: "The eternal awareness of choice can drive the weak to the point where the simplest decision becomes a nightmare. Most men prefer to flee choice, to flee anxiety, to flee freedom."[48]

Schlesinger's 1949 effort to move beyond the binary opposition of communism and capitalism contains within it a lesson about the decentralization of industrial society that is still potent today. Recalling the devotional aspect of the NPS's rhetoric suggested earlier in the paper, Schlesinger similarly traces the anxiety of the mid-century to the secularization of America through the consequences of industrialization and to the lack of a social replacement for religion. The abstraction of the new economic system had replaced the individual with the corporation but without providing a new social structure:

> Industrialism... imposed on the world a sinister new structure of relationships.... The result was to create problems of organization to which man has not risen and which threaten to engulf him; and it thereby multiplied man's anxieties. The result was to devitalize the old religions while producing nothing new capable of controlling pride and power; and it thereby heightened both guilt and anxiety.[49]

Addressing the failure of the right and the left in this regard, the possibility of a new radicalism in America can only draw its strength from the effort to secure the freedoms of the individual. Viewed as a form of political manifesto, the ideology of the NPS would seem to parallel Schlesinger's advocacy of a representational democracy.

For the National Park Service during the Cold War years, the problem of nature tourism was entangled with the presentation of the scenic view as a "found" and innocent situation. As James Corner notes in his useful conflation of the work of Dean McCannell and Michel Foucault, landscape—particularly when it is presented as benign and untouched—has the potential to obscure the ideological impulses that drive its formation and instead inspire in its occupants "the feeling that they are in possession of a beautiful and innocent past, that they have escaped from the inequities and problems of the present."[50] Relief from the anxieties of city life and, after the war, the cultivation of a normal, all-American lifestyle became an

Cover of a 1986 pamphlet describing the Man in Space initiative of the National Park Service

endemic part of the presentation of the landscape by the NPS. Through the efforts of Mission 66, the charge to "secure the beauty of America" entailed the orientation of the visitor within the park system and the enabling of individual experience and self-discovery. The particular typology of the visitor center allowed the task of orientation to find expression in civic form. At sites like Gettysburg, to this task was added the charge of attracting visitors to an otherwise featureless site. The practice of producing and inventing historical interest where none exists still motivates the NPS's interpretive efforts.

Public Law 96–344, passed in 1986, directed the Secretary of the Interior and the NPS to conduct a study of locations and events associated with the historical theme of man in space. Beginning with America's first flight in 1915 and concluding with its American voyage to the moon in 1972, the NPS proposed the preservation of twenty-five space-related landmarks—rocket engines, launch complexes, test facilities, wind tunnels, and mission control and tracking centers—through the creation of a national park. Because the sites were widely dispersed and in some cases inaccessible, an installation would need to be interpreted off-site, usually at visitor centers:

> The challenge is to provide the public with an overview of the Man in Space theme can be greatly enhanced by coordinating and expanding the focus of interpretive efforts beyond current and future space programs and by developing off-site movies, slide shows, displays, exhibits, publications, and active demonstrations.[51]

As an apotheosis of the site's detachment from the orientation center, inaugurated by early visitor centers, off-site orientation areas such as those of the Man in Space national park render the site itself unnecessary. Like the visitor centers pioneered during Mission 66, the success of Man in Space hinges upon the question of interpretation: building and view become superfluous.

BEAT SPACES

ROY KOZLOVSKY

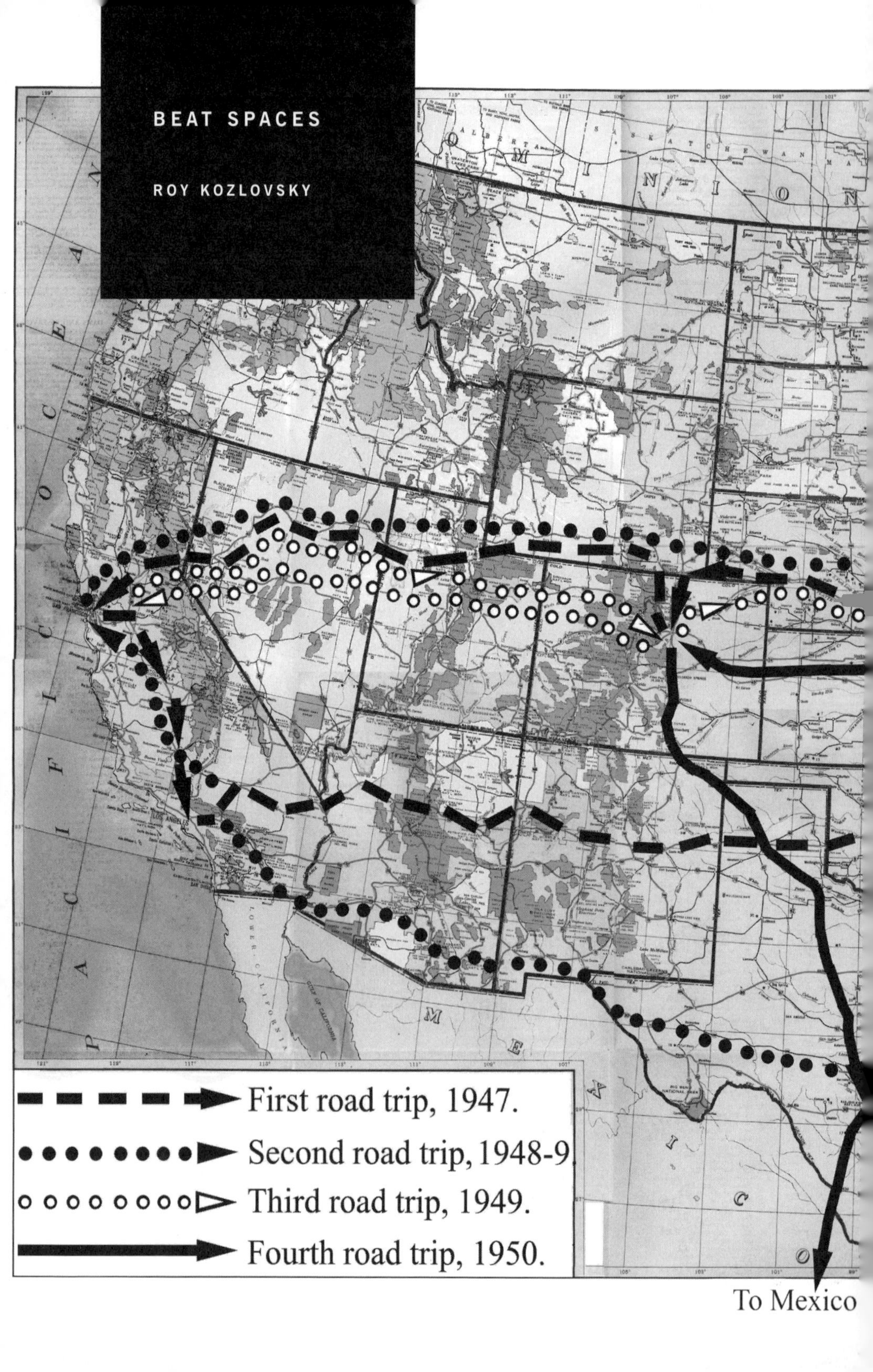

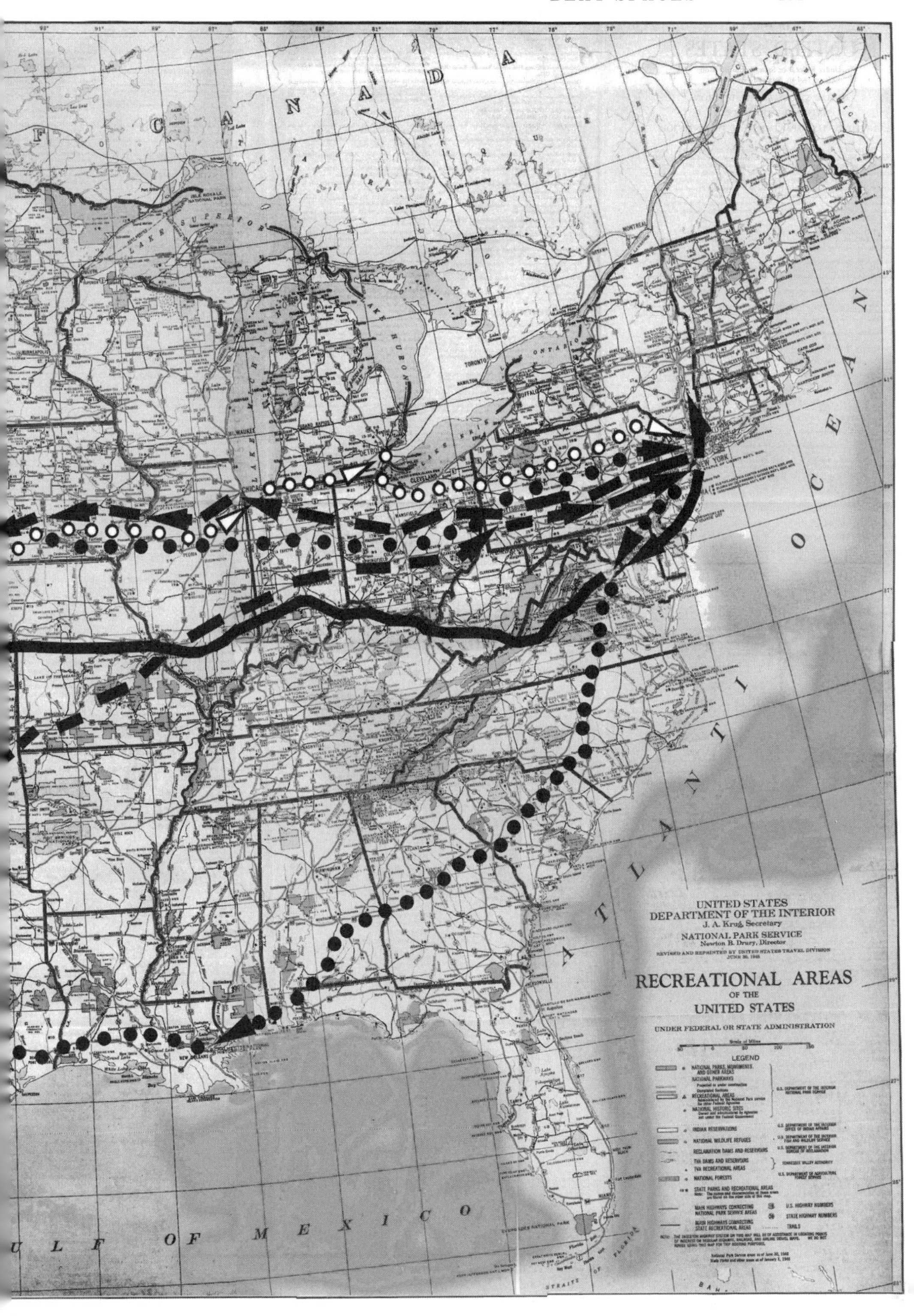
C A N A D A
A T L A N T I C O C E A N
U L F O F M E X I C O
UNITED STATES
DEPARTMENT OF THE INTERIOR
J. A. Krug, Secretary
NATIONAL PARK SERVICE
Newton B. Drury, Director
RECREATIONAL AREAS
OF THE
UNITED STATES
UNDER FEDERAL OR STATE ADMINISTRATION
LEGEND

The emergence of Beat literature in the 1950s was the outcome of a conscious attempt by a small group of writers to create a new American literary movement. These writers were acutely aware of inhabiting a cultural and political context radically different from that of their interwar literary predecessors, a novel context that necessitated a revision of the methods of writing, forms of narrative, and the social position of the writer. Their strategic choice was to transform literature into a spatial practice, as a means of exploring—and consequently critiquing—the emerging culture of post-World War II America. In doing so, they not only formed a literary movement but also constructed a new identity for their postwar generation.

The exploration and representation of American postwar space is the underlying theme of Jack Kerouac's *On the Road* (1957), the most popular work of the Beat movement. Upon its publication, the book was instantly recognized by the literary critic of *The New York Times*, Gilbert Millstein, as the first masterpiece of the "Beat Generation," a term which had been introduced into the cultural discourse five years earlier but still lacked a defining literary statement:

> "On the Road" is the most beautifully executed, the clearest and the most important utterance yet made by the generation Kerouac himself named years ago as "beat," and whose principal avatar he is. Just as, more than any other novel of the twenties, "The Sun Also Rises" came to be regarded as the testament of the "Lost Generation," so it seems certain that "On the Road" will come to be known as that of the "Beat Generation."[1]

On the Road had a lasting impact on American society, as it came to define an emerging youth culture that identified with its characters' spontaneity and affirmation of life, set in defiance of the established social norms of postwar America. This effect was not unintended by the author. Kerouac deliberately wrote in an innovative manner in order to differentiate his writing practice from those of his interwar literary predecessors, by transforming literature into a spatial practice.

The four road trips narrated in *On the Road* traced on a 1948 highway map of the United States (ove

The novel chronicles in detail four coast-to-coast road trips that the narrator, a character named Sal Paradise, and his companion Dean Moriarty make across the American continent over a period of four years. The work of fiction traces quite accurately the actual road trips taken by Kerouac, either alone or together with his friend Neal Cassady, who was the inspiration for the fictional character of Moriarty. But it would be misleading to regard the text as autobiographical in the sense that it chronicles an authentic lived experience; rather, the opposite: Kerouac decided to take these trips in order to invent his life so it could be made into literature. In a letter Kerouac wrote to his friend Hal Chase in April 1947, just before he began to live on the road, he disclosed the literary aims of this enterprise:

> I have begun a huge study of the face of America itself, acquiring maps (roadmaps) of every state in the USA, and before long not a river or mountain peak or bay or town or city will escape my attention. Now what does all this mean?...I know some people who would regard it as a step ahead. Because, after all, what is the ruling thought in the American temperament if it isn't a purposeful energetic search after useful knowledge. The "livelihood of man" in America instead of the vague and prosy "brotherhood of man" of Europe. My subject as a writer is of course America, and simply, I must know everything about it...Well, my purpose is Balzacian in scope—to conquer knowledge of the USA (the center of the world for me just as Paris was the center of the world for Balzac)—my purpose is to know it as I know the palm of my hand.[2]

Contrary to Balzac's panoramic literary project of mapping the social composition of nineteenth-century French society, Kerouac's research was less concerned with mapping a social reality or, for that matter, with "history"; his interest laid in geography and movement as an American alternative to the subject matter of European literature. Kerouac figured that, in the context of the postwar period, a writer who aspires to become the Balzac of American literature must physically explore the continental space of America. While on the road, Kerouac reported to his sister on the progress of his travels: "If I do this [go to Washington, Minnesota, etc.], and I'll certainly do 90% of it, I'll have seen 41 states in all. Is that enough for an American novelist?"[3]

By tracing the routes Kerouac traveled and later recorded in his novel, the political and cultural meaning of his project emerges. *On the Road* maps the concrete historical space as it was forged by the systematic transformation of American politics, economy, demography, and technology, which began with Roosevelt's New Deal and climaxed in the early Cold War era. It critiques specific developments in the structure and organization of American society that possess a significant spatial dimension: the integration of the United States into a unified continental system, the changing structure of urban life, and the redefinition of domesticity. The task of surveying this new space required the author to alter his mode of living and experiment with nomadic living practices. Fellow Beat writer William S. Burroughs observed that Kerouac had to develop an art of living in order to establish a new ethic of engagement between the writer and his work: "A whole migrant generation rose from *On the Road*. In order to write it the writer must go there and submit to conditions he may not have bargained for. He must take risks."[4] Such a choice was not unique to Kerouac. Other Beat writers, including Burroughs himself and Allen Ginsberg, spatialized their literary and living practices in order to write the other two seminal works of the Beat movement: the novel *Naked Lunch*, which Burroughs published in 1957, and the epic poem *Howl*, which Ginsberg published in 1956, two controversial texts that became the subject of censorship trials and consequently contributed to the struggle of the Civil Rights movement to defend and expand the freedom of speech.

As these three works demonstrate, Beat writers developed, as individuals and as a group, an engagement with space that informed their writing methods and literary production. Their aim was to carve out of the concrete space of America an alternative, fictional space that would redefine the relationship between space, society, and power. The Beat endeavor can thus be defined in retrospect as a *spatial project*, which can be recovered by overlaying the key work of Beat literature, Kerouac's *On the Road*, upon the political space of postwar America at its three distinctive scales—the continental, the urban, and the domestic.

CONTINENTAL SPACE

Tracing a work of literature on a map renders the relation between literature and space perceptible, since the map translates the text into a diagram that registers the tension between movement and place. Franco Moretti, in his study of the nineteenth-century European novel, employed such a method of mapping to argue that literary forms are place-bound: borders and geographical differences "give rise to a story, a plot."[5] Literature, and especially the genre of the novel, employs geography and movement in space to generate narrative and rhetorical tropes by engaging different regions, cultures, and social classes. When geography becomes suppressed or distorted, the ideological dimension of the novel is exposed.

The technical task of transcribing *On the Road* on a map of the United States is facilitated by Kerouac's insistence upon naming even the most obscure places that the narrator passes through, as if to allow the reader to follow his exact route. Some legs of the journey are meticulously marked with coordinates, while others are intentionally left in an untraceable state. By transcribing the road trips described in *On the Road* onto a highway map from the time in which the author began his cross-country trips, the spatial and geopolitical dimensions of the novel are brought to the fore.

What distinguishes *On the Road* from conventional travel journals is the cyclical movement of its main character, who often retraces previously taken paths while crisscrossing the continent. The pattern of movement displayed on the map raises several questions: What is the meaning of this repetition? What is the relationship between these lines and the social and political conditions of the 1950s? How do the different modes of transportation used by the characters affect the literary style of the text and its modes of subjectivity? In order to unpack the spatial configuration of *On the Road* on a continental scale, its engagement with space will be analyzed through the categories of organization, mobility, and mapping.

The map of the protagonist's journey records the spatial organization of the Beat movement within the United States. The lines on the map connect the places where Beat writers were physically situated during the 1950s—New York, New Orleans, Denver, San Francisco, and Mexico City. Often on the move and eventually dispersed across four continents (North and South America, North Africa, and Europe), Beat writers still managed to function as a group and collaborate on writing, editing, titling, and publishing their work. The correspondence that kept the Beats in contact had a direct impact on the poetics of their literary production. Burroughs incorporated his letters to Ginsberg and Kerouac directly into *Naked Lunch*. Kerouac developed *On the Road*'s innovative writing style, known as "spontaneous prose," from his correspondence with his friend Neal Cassady. Cassady's letters describing his sexual encounters while traveling the interstate bus system were written in a confessional and unedited style that Kerouac appropriated as his own.[6]

In the mid-1940s, before they dispersed, Beat writers were concentrated in New York City, near Columbia University, where Kerouac and Ginsberg were studying. The spatial pattern of Beat collaboration and communication, based on a network of concentration and dispersion, paralleled the spatial pattern of America in the postwar period. American space was reorganized after World War II through two complementary processes: it was decentralized by technologies such as the highway system and broadcast television, by accelerated suburbanization, and by the dispersion of military and civic means of production and research; at the same time, the production and management of the technologies and processes of dispersion became concentrated and centralized in a diminishing number of big corporations and government agencies.[7] Beat writers were in the position to experience these changes firsthand: Kerouac was employed during the war on the construction of the Pentagon, then the largest office building in the world[8]; Ginsberg worked in the advertisement industry and on the installment of the early warning system in the mid-1950s.[9]

The dual character of postwar space as composed of centrifugal and centripetal tendencies was exploited by Beat writers as a material base for their strategies of resistance and escape. The mobile and dispersed geography of Beat literary production was made possible by the continental networks that emerged during that period. The GI Bill—a social policy for funding higher education as well as providing the financial means for suburbanization—determined the yearly cycle of road trips chronicled in *On the Road*, since Kerouac depended on the periodic payments of his GI scholarship to finance his nomadic lifestyle. The national highway system, which was developed during World War II and became available to civilian exploration after the war ended, provided the medium for Kerouac's adventures, if not the very object of his literary investigation.[10] Kerouac was one of the first writers to transcribe the national highway network into literature, as a system that abolishes distance and unifies the American continent while at the same time decentralizing social space and individualizing its subjects.

The narrator of *On the Road* crosses the United States from east to west three times, and his final travel traverses the continent from north to south, reaching Mexico City. This pattern of movement was not the outcome of a spontaneous mode of living but was rather carefully planned. In the same letter to Hal Chase, Kerouac disclosed the literary sources for this spatial quest:

> Here's a list: Parkman's "Oregon Trail," another book concerned with that trail and also every other important trail in the country (don't ask me why: I'm crazy about this kind of reading now), a history of the United States, a biography of George Washington, a history of the Revolutionary War (campaigns and maps included).[11]

Kerouac explains the movement westward as an attempt to recover the meaning of the American dream by retracing its historical expansion. The first trip in *On the Road* partially follows the route described in *The Oregon Trail*: "We arrived at Council Bluffs at dawn; I looked out. All winter I'd been reading of the great wagon parties that held council there before hitting the Oregon and Santa Fe trails;

and of course now it was only cute suburban cottages of one damn kind and another."[12] The choice of the author to travel westward immediately after the war signifies an attempt to recover the concept of America by turning away from the country's origins in the East and moving in the historical and mythical direction of its open future. It is a search for becoming rather than a return to an origin.

The theme of direction is a recurrent feature of American literature. Leslie Fiedler argues in *The Return of the Vanishing American* that American literature can be classified according to its geography.[13] He identifies four archetypical types of literature; the Northern, the Eastern, the Southern, and the Western, each with its distinct content and imagery. The first three types are bounded to specific geographical regions, while the Western represents a process, or search, rather than any specific topography. Fiedler defines the movement westward as the symbolic escape of the male from the domain of the house and its "effeminate culture," as well as from the rule of the Father. It is allied with a quest for male companionship, replete with homoerotic undertones. Since the European imagines the American West as a territory uninhibited by "reason," the final attribute of the movement westward is madness.

Fiedler's literary compass offers a useful explanation for the turn westward as it is performed in *On the Road*. This movement is compressed into the character of Moriarty, who is portrayed as "Western, the west wind, an ode from the Plains, something new, long prophesied, long a-coming . . . a western kinsman of the sun."[14]

The move west allows the narrator to dispose of his old identity and reinvent himself. The road functions as a dissociative mechanism for the loosening of a fixed identity and as a vehicle for expanding experience. In the following passage, the narrator's subjectivity is literally transformed by the movement from east to west, as he becomes liberated from the burden of his past:

> I woke up [in Des Moines] as the sun was reddening; and that was the one distinct time in my life, the strangest moment of them all, when I didn't know who I was—I was far away from home, haunted and tired with travel. . . . I was just somebody else, some stranger, and my whole life was a haunted life, the

> life of a ghost. I was halfway across America, at the dividing line between the East of my youth and the West of my future, and maybe that's why it happened right there and then, that strange red afternoon.[15]

Beat writers opened another direction which Fiedler does not analyze: the movement southward. *On the Road* initially explores the American Deep South, but the last road trip continues further south toward Mexico City. While driving through Mexico, the narrator attempts to articulate the meaning of his southward movement:

> Not like driving across Carolina, or Texas, or Arizona, or Illinois; but like driving across the world and into the places where we would finally learn ourselves among the Felahin Indians of the world, the essential strain of the basic primitive, wailing humanity that stretches in a belt around the equatorial belly of the world from Malaya to India the great subcontinent to Arabia to Morocco to the selfsame deserts and jungles of Mexico.[16]

Mexican otherness is represented as the counterspace of Western civilization. The drive south is described as an escape from a Faustian culture of flux and negation, toward a rooted, harmonious civilization. Ironically, this attempt to escape Western modernity is contradicted by the very means of escape: the car—the vehicle of change and individuation—and the newly constructed Pan American highway. The utopian search for the American dream of companionship and union with nature is figured as a failure: "I realized I was beginning to cross and recross towns in America as though I were a traveling salesman—raggedy travelings, bad stock, rotten beans in the bottom of my bag of tricks, nobody buying."[17] The search for a "spontaneous," "natural," and "harmonious" society is unattainable, since it exists only as a counter-image to modern culture. Each road trip ends with the narrator returning to his aunt's home in Paterson, New Jersey.

The mobility of the Beat writers mirrors the concrete conditions of the war and postwar period, the most mobile decades in American demographic history. The war economy displaced as many as eighteen million Americans[18] and changed the settlement pattern of America by triggering a long-term migration from the Rustbelt to the Sunbelt and of Afro-Americans from the South to the Northeast and the West. In addition, demobilization of the armed forces after J-Day brought home in less than a year several million veterans.[19] These soldiers had lived through years of nomadic military life and did not necessarily return to their prewar homes. This resettlement is registered in the constant movement in space of the characters in *On the Road*.

The theme of direction is complemented by the concept of speed.[20] A central trope of the novel is the stimulating and liberating experience of accelerated driving. In the countless accounts of driving that compose the essence of the novel, speed is celebrated as a means for escaping the social in favor of the spatial: "'Whooee!' yelled Dean. 'Here we go!' And he hunched over the wheel and gunned her. . . . [W]e all realized we were leaving confusion and nonsense behind and performing our one and noble function of the time, *move*. And we moved!"[21]

The speeding up of movement in *On the Road* is tied to the construction of the American highway system. When the first motorized coast-to-coast trip was undertaken in 1903, it took as many as sixty days. By the time World War II was over, the highway system, built by the Defense Highway Act, had shortened the time to a few days.[22] The postwar conversion of the highway infrastructure to civilian use made it possible for individuals to explore speed and distance. The car democratized access to the experience of speed by allowing individualized control of speeding. It enhanced the shift from a passenger-oriented to a driver-oriented subject who identifies driving with individuality.[23]

The narration of *On the Road* makes a distinction between different modes of driving, with their corresponding speed, direction, and mode of subjectivity. The part of the journey that Paradise hitchhikes on his own is broken into short sections; its direction is westward, and its mode of subjectivity is dissociative and unstable. The trips he makes with Moriarty are composed of long and uninterrupted stretches of driving, since the two have a car under their

I first met met Neal not long after my father died. . I had just gotten over a
serious illness that I won't bother to talk about except that it really had so
thing to do with my father's death and my awful feeling that everything was
dead. With the coming of Neal there really began for me that part of my life
that you could call my life on the road. Prior to that I'd always dreamed of
West, seeing the country, always vaguely planning and never specifically ta
off and so on. Neal is the perfect guy for the road because he actually was
on the road, when his parents were passing through Salt Lake City in 1926,
jalopy, on their way to Los Angeles. First reports of Neal came to me th
Hal Chase, who'd shown me a few letters from him written in a Colorado reform
school. I was tremendously interested in these letters because they so naively
and sweetly asked for Hal to teach him all about Nietzsche and all the wonder
intellectual things that Hal was so justly famous for. At one point Allen
I talked about these letters and wondered if we would ever meet the strange
Cassady. This is all far back, when Neal wasn't the way he is tod
young jailkid shrouded in mystery. Then news came that Neal was
school and was coming to New York for the first time; also there
had just married a 16 year old girl called Louanne. One day I wa
the Columbia campus and Hal and Ed White told me Neal had just arrived
living in a guy called Bob Malkin's coldwater pad in East Harlem, the
Harlem. Neal had arrived the night before, the first time in NY, with
ful little sharp chick Louanne; they got off the Greyhound bus at 50 St. and
cut around the corner looking for a place to eat and went right in Hector's, and
since then Hector's cafeteria has always been a big symbol of NY for Neal. The

On the Road, original "scroll" manuscript, 1951

command; they travel westward, eastward, and southward, and their modes of subjectivity are distinguished by individuation and liberation. Although the longest in mileage, the sections in which the narrator travels by public transportation (typically the trips back home) are difficult to retrace, since few details are given. These trips signify a return to normalcy, as the protagonist is reduced by this means of mobility to the passive status of a consumer.

The new experience of driving provided by the highway system was employed by Kerouac to revolutionize the craft of writing. The author integrated the concepts of direction and speed into the technology of writing and literary style of *On the Road*. The initial drafts were written in the conventional, realist style of the travel novel. Kerouac then rewrote the entire text in twenty frenzied days; imitating Balzac, he wrote it on coffee.[24] He typed directly on several rolls of architectural drafting paper, which he taped together into a long scroll 9 inches wide and 119 feet, 8 inches long.[25] The manuscript of *On the Road* assumes the form of a straight, unfolding road in order to transfer the experience of movement and speed into the mechanics of writing. The process of typing on the scroll, invented by Kerouac to overcome the interruption caused by loading individual sheets of paper into the typewriter, corresponds to the experience of uninterrupted driving on a highway. The act of typing, like driving, becomes a mode of spontaneous expression, uniting man and machine. Such a method of writing forecloses the possibility of editing or revising the original and authentic act of writing. Kerouac named this type of to literary production "spontaneous prose."

The experience of driving is incorporated into the language of the text in the form of metaphors. Writing becomes a metaphor for driving, and the driving experience is likened to a textual operation: "The magnificent car made the wind roar; it made the plains unfold like a roll of paper.... I opened my eyes to a fanning dawn; we were hurling up to it. Dean's rocky dogged face as ever bent over the dashlight with a bony purpose of its own."[26] The scroll format changes the relationship between the linguistic and material aspects of a text: the physical object of the scroll has a presence in the language of the text.

The consolidation of the United States into a unified and interconnected space made it possible to think and write about the country as a totality. In the context of postwar America, the ways in which this territorial unity would be conceptualized had significant political implications. Two different mapping projects undertaken in the late 1940s and early 1950s, by Senator Joseph R. McCarthy and the sociologist Alfred C. Kinsey, respectively, demonstrate that mapping was a contested political operation in which conflicting bio-political approaches collided.

McCarthy employed maps to visualize the spread of communism within the United States as part of his crusade against anti-American activity. His map is more than a rhetorical device for exaggerating

Senator Joseph R. McCarthy, testifying at the House Committee on Un-American Activities, aided by a map surveying communist presence in America, 1950

the threat of communism; it is an instrument of control that is a product of a regime of surveillance. It creates a powerful analogy between the nation and the body. At the beginning of the Cold War, the boundaries between America and the external threat of communism were redrawn internally. McCarthy constructed an equation between political and sexual deviation whereby homosexuality was linked to political subversion and penetration from the outside.[27] He led a campaign to purge the federal government of employees suspected of sexual deviance in order to protect the national body from "contamination" from within. Beat writers rewrote and subverted this analogy for their own sexual-political aims. In the poem *Howl*, for example, Ginsberg describes a scene in which he is in bed with a sick America. The scene takes place in the Rockland State Hospital, a psychiatric institution to which the young writer was confined in 1947 to "cure" his homosexuality (or face expulsion from Columbia University):

> I'm with you in Rockland
> where we hug and kiss the United States under our
> bedsheets the United States that coughs all night and won't
> let us sleep.[28]

The body metaphor is mocked and reversed by Ginsberg—America is laid in bed with a homosexual.

The metaphor of America as a body is employed by Kerouac in a different manner: "And here for the first time in my life I saw my beloved Mississippi River, dry in the summer haze, low water, with its big rank smell that smells like the raw body of America itself because it washes it up."[29] More erotic than critical, Kerouac's metaphor suggests a conception of *On the Road* as a performance piece in which the author aims to establish an intimate knowledge of America by physically mapping its surface. In his mapping activity and his employment of the America-as-body metaphor, Kerouac also explores marginal spaces, extracted, so to speak, from the national body: "[Laredo, Texas] was the bottom and dregs of America where all the heavy villains sink, where disoriented people have to go to be near a specific elsewhere they can slip into unnoticed."[30]

The literary device of mapping not only supported the analogy of America as a bodily unit but also permitted a counterstrategy of its fragmentation as a critique of power. As the Kinsey survey of male sexual behavior from 1948 demonstrates, mapping could yield a radical critique of society by denaturalizing its most fixed and revered conceptions of the body and of sexuality. Trained as a zoologist, Kinsey mapped the sexual behavior of males nationwide from a factual, scientific point of view. He found that the realities of sexual activity radically differed from the moral conventions and popular conceptions espoused by the public and its legislators. Kinsey determined that as many as half of the male population had had some kind of homosexual experience, which was illegal in the 1950s. As a result of his findings, he made the claim that human sexuality was not a fixed biological category that could be divided into normal and pathological manifestations but rather was flexible and fluid, as sexual desire could be attached to different objects and genders and performed in multiple ways. Kinsey's method of mapping allowed different, even incompatible, cultural identities to be represented side by side. In the hands of scientists such as Kinsey, mapping became an epistemological tool used to represent, in a non-hierarchical and inclusive manner, what society wished to exclude or marginalize. It allowed one to deconstruct and historicize dominant perceptions of sexuality that had been presented as natural and then imposed by force:

> Conflicts between social levels are as intense as the conflicts between nations, between cultures, between races, and between the most extreme of the religious groups. The existence of the conflict between sexual patterns is, however,

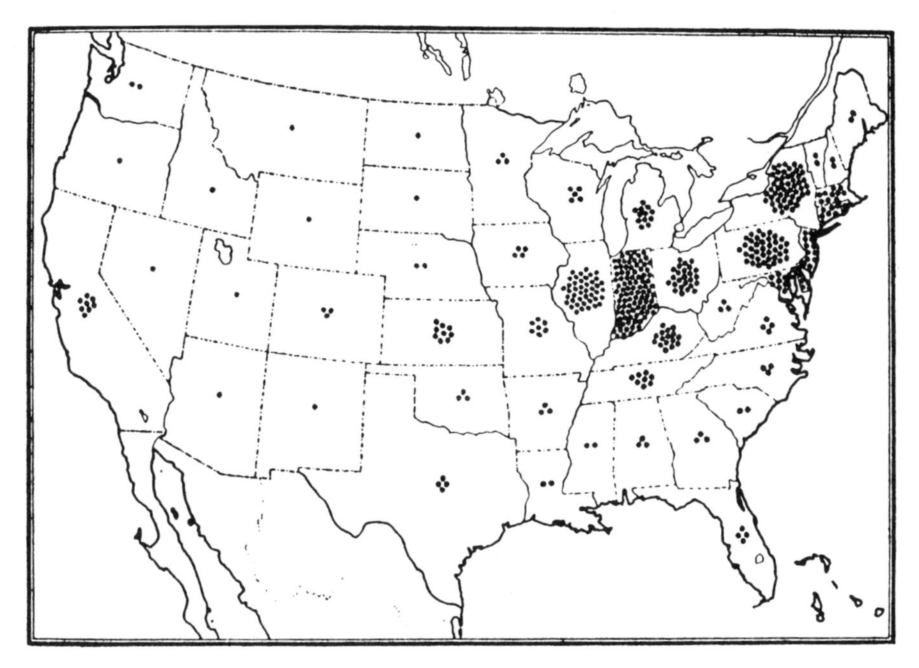

Map showing the geographical distribution of Kinsey's survey of male sexuality. One dot represents fifty men surveyed.

> not recognized by the parties immediately concerned, because neither of them understands the diversity of patterns which exist at different social levels.[31]

The Kinsey survey forms a backdrop for the action in *On the Road*. Beat members had participated in the survey and it is mentioned by Kerouac's character in a number of instances. Its presence illuminates the political intention behind the Beat practice of sexualizing literature and of using mapping techniques to accommodate difference.

Mapping also allows different modes of language to exist side by side. The different idioms and vernaculars that Beat writers collected in their geographical surveys were incorporated in their texts, producing a carnivalesque effect that undermined the hegemonic status of the dominant cultural idiom. *On the Road* invites "high" and "low," Shakespearian and vernacular forms of English, to coexist, forming a textual mosaic. A typical example is provided by Carlo Marx, the fictional character fashioned after Allen Ginsberg, who remarks, "What is the meaning of this voyage to New York? What kind of sordid business are you on now? I mean, man, whither goest thou? Whither goest thou, America, in thy shiny car in the night?"[32]

Mapping is employed by Kerouac as a form of resistance. At the historical moment when the state was attempting to homogenize and normalize American society and to impose the ideology of containment and exclusion from without as well as within, the mapping of the linguistic and cultural surface of American culture by the Beats opened up an alternative space of multiplicity and heterogeneity. This counter-space challenged the rigid boundaries between high and low, normal and pathological, interiority and exteriority that the postwar order tried to enforce.

URBAN SPACE

On the Road chronicles thousands of miles of driving back and forth across the continent in straight lines that connect major cities, for the most part bypassing the space in-between:

> We had come from Denver to Chicago via Ed Wall's ranch, 1180 miles, in exactly seventeen hours, not counting the two hours in the ditch and three at the ranch, and the two with the police in Newton, Iowa, for a mean average of seventy miles per hour across the land, with one driver.... Great Chicago glowed red before our eyes.[33]

As this passage exemplifies, the geography of *On the Road* is split between movement on the highways and life in the cities, and as much as it is a travel novel, it is equally a representation of an urban culture that is situated in a specific historical context. In the postwar period, the long-term process by which both rural and urban populations migrated to the suburbs of metropolitan regions was accelerated to the point that the American city lost its primacy as the dominant living environment. By 1950 the suburbs were expanding ten times faster than the cities in order to accommodate the "baby boom."[34] The populations of eighteen out of the twenty-five largest American cities in 1950 would decrease in the following decades.[35] The inner city was represented in mass media as the counter-image of the suburb—a dangerous, crime-ridden, and poverty stricken zone. In this historical context, the work of Beat writers cuts against the grain of the great spatial transformation of American urban space, as it made the city the privileged site of its poetics. Like many other critics, Beat writers rejected the suburban ideal, pointing to the realities that the suburbs suppressed. Burroughs' condemnation of suburbia is typical: "It's like I can't breathe in the U.S., especially in suburban communities. Palm Beach is a real horror. No slums, no dirt, no poverty. God what a fate to live there!"[36]

It would be erroneous, however, to represent the relation of the Beats to the city in terms of the urban-suburb discourse, as their literary interests lay elsewhere. The Beat movement was foremost an urban literary movement. Its proponents rejected the autonomy of the academy in favor of the street, making a strategic choice to introduce their art into the public sphere. They incorporated the experience of the city into their poetics and explored the city's marginal populations and spaces in order to express and affirm the urban conditions of disorder and heterogeneity. Beat writers were interested in cultures and modes of experience that escaped the process of normalization. They became familiar with marginality by embarking on a project to live in borderline urban spaces and to associate with criminals, drug addicts, ex-convicts, and oppressed racial groups.

Although Kerouac, Ginsberg, and Burroughs, as well as other Beat writers, were incarcerated for various offenses, their conflicts with the law were not ideologically motivated, as compared to writers who were imprisoned or blacklisted for their beliefs during the Red Scare of the early 1950s. Their incarcerations were more a product of their art of living, which combined an anthropological method of observation with a commitment to unmediated bodily experience of the city. In *On the Road*, Kerouac describes how Burroughs (represented in the novel by the character of Old Bull Lee), who studied anthropology at Harvard

University, was gaining knowledge of the "facts of life" by practicing an independent type of anthropological research: "He dragged his long, thin body around the entire United States and most of Europe and North Africa. . . . He was an exterminator in Chicago, a bartender in New York, a summons-server in Newark. . . . He did all these things merely for the experience. Now the final study was the drug habit."[37]

Positioning themselves outside the norms of society, each Beat writer explored a different social strata of the city. Kerouac was foremost interested in black urban culture, and especially in jazz. The cities that Paradise visits in *On the Road* were the centers of bebop activity when the genre was still an urban affair, not yet incorporated into the music industry. The jazz scenes in the novel are frequently hailed by critics as the most poetic commentary written on jazz by an American writer.[38] What is often overlooked, however, is the way in which jazz is represented as an urban event, spilling from the stage to the public space of the street. The narrator and his companion Moriarty often join black musicians in a spree of drinking and reckless driving.

Kerouac's ideal was to incorporate jazz into his poetics; in his manual for writing Beat literature, *Essentials of Spontaneous Prose*, he appropriates jazz as the model for the styling of language and for the act of writing itself:

> METHOD No periods separating sentence-structures . . . but the vigorous space dash separating rhetorical breathing (as jazz musician drawing breath between outblown phrases)
> CENTER OF INTEREST Never afterthink to "improve" or defray impressions, as, the best writing is always the most painful personal wrung-out tossed from cradle warm protective mind—tap from yourself the song of yourself, *blow!—now!—your* way is your only way.[39]

The incorporation of the jazz ethos into literature was a contested issue in the 1950s, since jazz was regarded by the white elite as a primitive art form. In addition to its literary dimension, the turn to jazz had a political significance, as the affirmation of Black music within mainstream American culture contributed to the long process of breaking down the color barrier.

In *On the Road*, Kerouac explores the spatial dimension of racial discrimination in America, in which people of color were segregated into separate zones of the city. The first road trip ends in Mill City (present-day Marin City), one of the few public housing projects built during World War II that was not segregated according to race:

> Mill City, where Remi lived, was a collection of shacks in a valley, housing-project shacks built for Navy Yard workers during the war.... It was, so they say, the only community in America where whites and Negroes lived together voluntarily; and that was so, and so wild and joyous a place I've never seen since.[40]

Drawn to the subjective experience of the city, Beat writers experimented with various techniques of exploring urban space and incorporating urban subjectivity into their textual operations. Their practices resemble those developed by the surrealists and later by the situationists: drifting without a destination, experiencing urban space under the influence of drugs, exploring spaces that escaped rationalization and modernization. If the situationists idolized the Les Halles market in Paris and made it into the epicenter of their urban drifting activity,[41] the Beats were equally attracted to Times Square. All the road trips in *On the Road* end in New York, and more precisely, in Times Square: "Suddenly I found myself on Times Square. I had traveled eight thousand miles around the American continent and I was back in Times Square."[42] Times Square informed the Beat's conception of the city as a heterogeneous, porous space, not yet rationalized and organized into separated, normalized functions. It was also where Beat writers made their connections with underworld characters trafficking in sex, drugs, and stolen goods:

> Ritzy's bar is the hoodlum bar of the streets around Times Square.... There were wild Negro queers, sullen guys with guns, shiv-packing seamen, thin, non-committal junkies.... Kinsey spent a lot of time in Ritzy's bar, interviewing some of the boys; I was there the night his assistant came, in 1945. Hassel and Carlo were interviewed.[43]

The reference to Kinsey and his survey of male sexuality in the context of Times Square suggests a conception of space that allows a multiplicity of practices and a fluidity of subjectivities to be experienced side by side. It marks Kerouac's opposition to the tendency of postwar power to force a fixed and legible norm upon the subject.[44]

Although it was not the explicit agenda of Kerouac or the Beat writers, the discovery of the complexity and plurality of urban experience prefigured the rediscovery of the city by urban theorists such as Kevin Lynch and Jane Jacobs. If Lynch's project was to reclaim the lost identity and legibility of the city as a visual and cognitive experience, and if Jacobs's was to revive traditional values of community and street life as an aesthetic experience, the Beat urban project was to explore the postwar conditions of urban disorder and inequality. Kerouac made those aspects of the city that were considered abnormal into the foundation of his poetics and politics.

DOMESTIC SPACE

After World War II, American society was restructured between two spatial poles—the continental and the domestic. The GI Bill, which financed mass homeownership and education as means of social mobility and stability, established the material base for organizing American everyday life and mass culture on the family unit. The home—the site of sexual reproduction and material consumption—was promoted as the moral and economic ideal of the Cold War era. The remark of the pioneer of postwar suburban developer William Levitt that "No man who owns his own house and lot can be a Communist"[45] reveals the ideological dimension of domesticity in the late 1940s: homeownership as an antidote to the menace of socialism. Deviators from this domestic ideal were labeled un-American. In the context of this concrete historical and spatial process by which American society was organized into the confines of the nuclear family, Beat writers explored in their artistic and living practices radical alternatives of communality.

On the Road offers a critique of the domestic domain on several levels. It views the conditions of homelessness and mobility as alternative practices to settling down in a suburban home, with all its implied responsibilities and gender roles. Heterosexual domestic space is represented in negative terms, as a site of perpetual crisis where the two genders argue constantly over the male's unwillingness to capitulate to the demands of the domestic economy. Ironically, the only portrait of a functioning domesticity in *On the Road* is of Old Bull Lee's family, a matrimonial union between a homosexual husband with a heroin habit and his Benzedrine-addicted wife. They live on a ranch with an organon in the backyard and farm marijuana for the New York market, yet they are portrayed as the tender parents of two angelic children and the only couple in the novel who have a reciprocal relationship.

As an alternative to the realm of normative domesticity, *On the Road* proposes an ethics based on mobility. In a recurring scene in the novel, the nomadic protagonists leave behind their settled friends:

> We roared off. We left Tim in his yard on the Plains outside town and I looked back to watch Tim Gray recede on the plain.... He grew smaller and smaller, and still he stood motionless with one hand on a washline, like a captain, and I was twisted around to see more of Tim Gray till there was nothing but a growing absence in space, and the space was the eastward view toward Kansas that led all the way back to my home in Atlantis.[46]

Home as the place of origin exists only as the mythical, lost space of Atlantis. Kerouac likens homes to marooned ships. Homelessness—a historical condition brought on by the economic legacy of the Great Depression, with its subculture of migrant workers, hobos, and Okies—is elevated by the author to a condition of choice. Mobility is represented as a productive practice that generates new forms of sociability and communality that are embodied in the ideal of "beatness" and its rejection of the cycle of work and consumption in favor of the immanence of life:

> Although [Mississippi] Gene was white there was something of the wise and tired old Negro in him, and something very much like Elmer Hassel, the New York dope addict, in him, but a railroad Hassel, a traveling epic Hassel, crossing and

> recrossing the country every year, south in the winter and north in the summer, and only because he had no place he could stay in without getting tired of it and because there was nowhere to go but everywhere, keep rolling under the stars, generally the Western stars.[47]

If the vehicles of the earliest novel, *Don Quixote*, are a horse and a mule, and the vehicle of the nineteenth-century novel is the boat, as exemplified by Conrad, Melville, and Twain, then the vehicle of the twentieth-century novel is the car. Kerouac poses the car as an alternative space to the house, the real intimate space of *On the Road* where male companionship can be reestablished in the face of "feminine" domestication. The car functions as a communication machine that assumes the role of the hearth in the construction of communality:

> Stan swung into his life story as we shot across the dark.... Suddenly we passed Trinidad [Colorado] where Chad King was somewhere off the road in front of campfire with perhaps a handful of anthropologists and as of yore he too was telling his life story and never dreaming we were passing at the exact moment on the highway, headed for Mexico, telling our own stories.[48]

At the time that Kerouac was writing *On the Road*, the effects of mass car ownership were the subject of critical debate. The sociologist David Riesman argues in his 1956 essay "Autos in America" that the car changed from a utilitarian machine with a positive use value into a commodity fetish designed by market research to satisfy false desires created by advertisement. He condemns this cycle of production and consumption as socially wasteful and equates it with the logic of the arms race with the Soviet Union, an element of a military-Keynesian economy addicted to producing waste in order to maintain economic stability.[49] In the context of Riesman's critique of the economy of the car, Kerouac suggests that the car should be destroyed in a potlatch fashion. In *On the Road*, private ownership of the car is made insignificant, as much of the trip is done by bus, by sharing car ownership, by hitchhiking, and by the lending or stealing of vehicles. The aim of driving is to push the machine to the point of mechanical breakdown, until the car is destroyed. Kerouac underscores the waste and violence of the capitalist system by subjecting the "conspicuous" Cadillac, the "functional" Ford and the "feminine" Plymouth—distinctions that correspond to the different production ideologies of cars as outlined by Riesman—to the same destructive end.[50] Kerouac turns the car, which was advertised during the 1950s as a domestic machine for the heterosexual family, into its adversary—a temptation for the male to betray his domestic responsibilities:

Neal Cassady and Jack Kerouac, San Francisco, 1952

> [Dean Moriarty] became the father of a cute little girl, Amy Moriarty. Then suddenly he blew his top while walking down the street one day. He saw a '49 Hudson for sale and rushed to the bank for his entire roll. He bought the car on the spot.... Now they were broke. Dean calmed Camille's fears and told her he'd be back in a month. "I'm going to New York and bring Sal back." She wasn't too pleased at this prospect.[51]

The ideal of domesticity was regarded by the Beats as an imposition of middle-class values, suppressing other possible forms of sexuality and experience. Against the normative heterosexual family unit, Kerouac posited a more fluid and dynamic mode of habitation. In *On the Road*, Moriarty hops between the beds of his wife

Camille at home, his ex-wife Marylou at a hotel, and Carlo Marx in his basement flat, all in one day—a feat that appears plausible in the context of Kinsey's report on the sexual practices of American "lower classes," to which Moriarty belongs. Intercourse with both male and female partners several times in one day was not regarded as unnatural or immoral among less-educated populations, but it was Kerouac's portrayal of this behavior that scandalized *On the Road* for its critics.[52]

There is a blunt misogynist element to Beat writing, in which the female gender is reduced either to the status of an object of male desire or to the role of a castrating force determined to lock up the male in a domestic prison. It is in this reactionary and anti-feminist capacity that *On the Road* celebrates as a domestic ideal a scene in which Moriarty and Paradise pay a late-night visit to the home of a black jazz musician. His wife obediently serves the husband and his companions, causing Moriarty to reflect: "Now you see, man, there's *real* woman for you.... [T]his is a man and that's his castle."[53] Beat misogyny can be read as the rearguard reaction of the male to the change in gender roles in American society following World War II, when women were first mobilized on a massive scale in the workforce, upsetting the domestic division of labor and improving their social status.[54]

CONCLUSION

The Beat project of transforming literature into a spatial practice corresponds with Frederic Jameson's assertion in *Postmodernism, or, The Cultural Logic of Late Capitalism* that the shift from "high modernism" to "post modernism," which he traces to the 1950s, represents the shift from the modernist parameters of time and history to the postmodernist parameters of surface and space:

> We have often been told...that we now inhabit the synchronic rather than the diachronic, and I think it is at least empirically arguable that our daily life, our psychic experience, our cultural languages, are today dominated by categories of space rather than by categories of time, as in the preceding period of high modernism.[55]

Jameson is critical of the spatialization of cultural production, arguing that the suppression of the historical in favor of the spatial is symptomatic of the multinational phase of late capitalism and the subsequent decline of a class-based struggle. The case of Beat literature might suggest that the spatialization of literature is not merely symptomatic but served a critical agenda. If the Beat's contemporary critics have generally conceived Beat literature as apolitical,[56] it is only because the definition of what was political was changing exactly during that period, when critical discourse shifted from the conflict between mass ideologies of production to one about power. The American political struggles of the 1950s were concerned with racial, sexual, and gender emancipation and were fought in the sphere of individual rights. Their central concern was the attempt of the modern state, regardless of its ideological position, to homogenize and normalize its heterogeneous populations into one rational economic system. In such a constellation, Beat literature was political indeed, as its insistence on reconfiguring the boundaries between private and public was aimed at critiquing exactly those sites of social conflict and oppression that were kept out of the public discourse by being confined to the domestic realm. Such a position corresponded with the realities of postwar America, in which power was exercised through new technologies and modes of organization that were essentially about how social space was divided and controlled.

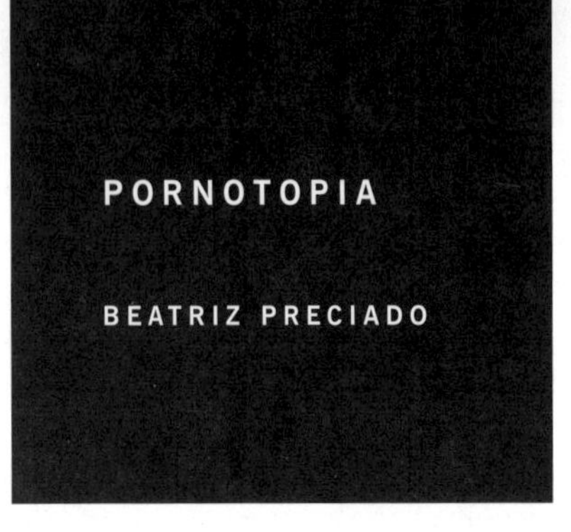

PORNOTOPIA

BEATRIZ PRECIADO

A public space is not a space in itself but the representation of a space.
—Vito Acconci

In 1962, Hugh Hefner was photographed posing as an architect, exactly as Le Corbusier or Ludwig Mies van der Rohe might have been photographed earlier in the century. Indifferent to the camera, avoiding any relationship to the spectator, his eyes establish a privileged connection to the architectural model next to which he stands. His body is turned toward it, and his hands point reverently to details of construction as if he were tied to the building through the bonds of creation. There is nothing strange in Hefner's pose, except that he was not the building's architect but rather the creator of *Playboy*, the first mainstream pornographic magazine in America; the mock-up, a model for a new Playboy Club Hotel to be built in Los Angeles.

Far from farcical, Hefner's posing as an architect speaks to the very heart of the production of his so-called "Playboy Empire." The first issue of *Playboy* magazine was produced in 1953 on the kitchen table of Hefner's South Side apartment in Chicago. Hitting newsstands in December of that year, it carried no cover date because its publisher was not sure when or if he would be able to produce another. The issue sold more than 50,000 copies—enough to cover its costs and finance another issue. From that point on, *Playboy* magazine

From Griffin, Diane
Sent Wednesday, January 8, 2003 4:06 pm
To Brennan, AnnMarie
Subject Request to use PB images in the book "Cold War Hothouse"
re: Your request to reprint 17 Playboy images within the book "Cold War Hothouse"

Dear Dr. Brennan:

Company policy in regard to the release of our material to third parties is as follows:
(1) We never align ourselves with the word "pornography." Unless the chapter title "Pornotopia" and the references to pornography are removed, we will not release any material.
(2) We only allow up to five images (art/photo/cartoons/text) for use in one volume of work. Note that some of the illustrations you selected feature multiple images--photos, text and art. For example, your illustration numbered 4 features five photos, one illustration and text.
(3) We do not release any photos featuring nudity (your illustrations numbered 3, 5, 13, 16).

Let me know if you want to proceed with this request.

Sincerely,
Diane Griffin
Rights and Permissions Records Manager
Playboy Enterprises International, Inc.

NOTE FROM THE AUTHOR:
Although originally planned as an illustrated piece, this article appears here unillustrated. Playboy Enterprises International, speaking through its Rights Manager, made as a condition of the release of any visual material under their copyright the removal of the word "pornography" from my text. Taking into account the aim of my inquiry—to investigate the relationship between architecture and pornography as a visual regime during the 1950s—I decided to preserve the integrity of the piece and publish it without images, rather than heed Playboy's recommendation. This was both a theoretical and political choice on my part. Somewhat paradoxically, the absence of images in this article confirms my thesis about the operation of pornography as a multimedia architecture, the power of which resides in its ability to redefine the limits of publicity and privacy and to control the processes through which spaces, images, words, and bodies are rendered visible or invisible. —B.P.

took the form of an expansive enterprise, directed toward the construction of the multi-media Playboy House.

During the 1950s and early 1960s, Hefner created some of the most famous interior spaces in America: first, his own house in Chicago, known as the Playboy House, and later, more than fifteen Playboy Clubs across the country. With regard to his seemingly domestic project in Chicago, Hefner explained:

> I wanted the house to be a dream-house. A place where one could work and have fun without the trouble and conflicts of the outside world. Inside, a single man had absolute control over his environment. I could change night into day screening a film at midnight and ordering a dinner at noon, having appointments in the middle of the night and romantic encounters in the afternoon. It was a haven and a sanctuary.... [W]hile the rest of the world seemed to be out of control, everything inside the Playboy House was perfect. That was my plan. Being brought up in a very repressive and conformist manner, I created a universe of my own where I was free to live and love in a way that most people can only dream about.[1]

In *Fast Cars, Clean Bodies: Decolonization and the Reordering of French Culture*, critic Kristin Ross calls "privatisation," the movement inward that took place after World War II,

> [the] withdrawal of the new middle classes to their comfortable domestic interiors, to the electric kitchens, to the enclosure of the private automobiles, to the interior of the new vision of conjugality and an ideology of happiness built around the new unit of middle-class consumption, the couple, and to de-politization as a response to the increase in bureaucratic control of daily life.[2]

Although *Playboy*'s efforts to reformulate interior space could be taken as part of this process of privatization, its aims and strategies paradoxically share little in common with this notion of the private. The Playboy House represents instead an extreme alternative to the single-family home of the 1950s, constructing a possibility for domestic space to function as "public."

If Hefner enjoyed portraying himself as an architect, it is precisely because he perceived his role as creator of the Playboy Empire (of which *Playboy* magazine was nothing but the paper organ) as part of a larger discourse that promoted a new anti-domestic and yet interior regime, or even a form of non-domestic interiority. His enterprise could be described as a sort of minor architectural project that would enable him to produce "a world inside another world," to develop a new, male, domestic interior where one could, according to Hefner, enjoy the privileges of public space without being subjected to its laws, and dangers.[3]

The spatial economy described in *Playboy* magazine and inscribed within the Playboy House between 1953 and 1963 articulates itself around oppositions, such as inside/outside, private/public, work/leisure, dressed/undressed, one/many, dry/wet, human/animal, controlled/relaxed, fidelity/promiscuity, vertical/horizontal, and family/stranger.[4] Reluctant to take a position among these radical oppositions, the Playboy appeared as a liminal subject whose final decision was just to "play." This "play" materialized in the use of what can be identify as *dispositifs* of rotation that stressed flexibility, reversibility, and circularity. These enabled the constant conversion of one term of a given opposition into the other. The 1956 Playboy Penthouse Apartment, as well as the 1959 Playboy House, employed a certain number of objects, visual devices, and architectural apparatuses to convert work into leisure, dressed into undressed, one into many, and so on, and vice versa.

Three circumstances inscribed the "play" of these *dispositifs*. First, the actor—the only one entitled to play—is the male reader-client (and ultimately) TV-viewer for whom the seduction rhetoric is constructed.[5] Second, it is basic to the functioning of these *dispositifs* that they be reversible, so that any changes are local and temporal rather than general and permanent. And third, this switching activity, this *Fort-Da* game, produces pleasure (fun, entertainment, amusement, etc.), the highest form of which was sexual pleasure, the "play" that gives the name to the magazine.

THE RISE OF MALE DOMESTIC AWARENESS AND THE SEARCH FOR A "QUARTERS" OF ONE'S OWN

In the recent book *Inside the Playboy Mansion*, commissioned by Hugh Hefner, journalist Gretchen Edgren reconstructs Hefner's biography and *Playboy* magazine's story through more than one thousand images of the interior of his houses: the Playboy House (1959) in Chicago and the Playboy Mansion West (1972) in Los Angeles. The people that entered in, the games that were played inside, how the magazine was produced right from Hefner's famous rotating bed, the TV programs shot there, the Jazz festivals and "staged" parties that were held inside: all are documented by Edgren. All told, after 1959 Hefner spent more than forty years without leaving home, wearing only his pyjamas and slippers.[6] The book relies on a single narrative that explains the emergence of the magazine as part of one and the same process of house construction. In fact, its rhetorical argument seems to suggest that *Playboy* magazine contributed during the 1950s, through the promotion of various penthouse ideals, to the emergence of what is paradoxically described as a political male domestic awareness.

From 1953 to 1963, *Playboy* provided a new discourse for the production of the young American urban bachelor. The new unmarried male proved to be the central figure within the first successful counter-narrative to the American dream. Against the "heaven of the heterosexual family house," *Playboy* argued for the construction of a parallel utopia, a "haven for the bachelor in town."[7] It is in this spirit that in 1953, Hefner would define *Playboy* in the editorial of the first issue as an "indoors magazine":

> Most of today's 'magazines for men' spend all their time out-of-doors thrashing thorny thickets or splashing about in fast flowing streams. We'll be out there too, occasionally, but we don't mind telling you in advance—we plan on spending most of our time inside. We like our apartment.[8]

In fact, Hefner launched the first number of *Playboy* magazine in November 1953 from his own apartment with only the help of his first wife Millie Williams and some friends.[9] Moreover, the editorial address printed inside the magazine was the young Hefner's family domicile.[10]

Today, the Web page dedicated to Hugh Hefner at salon.com portrays this "interiorization" as one of *Playboy's* contributions to the refashioning of masculinity in American culture:

> *Playboy* brought men indoors. It made it OK for boys to stay *inside* and play. Where other men's magazines—*Argosy*, *Field & Stream*, *True*—affirmed their readers' places in duck blinds and trout streams, Hef's took men inside to mix drinks, sit by the fire and play backgammon or neck with a girlfriend. In what would later become an ironic collusion with feminists such as Betty Friedan, *Playboy* critiqued the staid institutions of marriage, domesticity and suburban family life. Suddenly bachelorhood was a choice, one decorated with intelligent drinks, hi-fis and an urbane apartment that put white picket fences to shame. Sophistication had become a viable option for men: The *Playboy* universe encouraged appreciation of "the finer things"—literature, a good pipe, a cashmere pullover, a beautiful lady. America was seeing the advent of the urban single male who, lest his subversive departure from domestic norms suggest homosexuality, was now enjoying new photos of nude women every month.[11]

Against the rigid gender division of spaces of the pre-World War II period that defined public, exterior, and political domains as masculine battlefields and private, interior, and domestic spaces as naturally suited to women, *Playboy* promoted the masculine occupation of interior space. This was not the first time that men initiated a homecoming; just ten years earlier, American soldiers had returned home after fighting the dangers of the outside (in the double sense of outside the home and outside the country). Whereas soldiers had proved their masculine power fighting against the dangers of the outside space, they had simultaneously lost control over the precious space of the home. The return of the male to the interior space was presented by *Playboy* as a form of active compensation, of supplementing an excess of masculinity that would have led the traditional American male to neglect the details of the interior. But this second homecoming was more like a moving: the new male, at once unsuited for monogamous family life and scientifically aware of the radiating dangers on the exterior, stepped back into the house as woman's strongest rival—rather than complementary partner—promoting a new gender segregation within the domestic realm.[12]

As part of this "penetrating" agenda, from 1953 onward almost every *Playboy* issue included an article on the reappropriation by the bachelor of interior, quasi-domestic space: the glamorous weekend house in the country, the private yacht, the studio, or simply the car. This program of recolonization reached its climax in the reportage about the Playboy Penthouse Apartment, published in September and October 1956. The colored sketches of the penthouse design were most likely inspired by the apartment of Victor A. Lownes, Hefner's associate during the 1950s, but reflected Hefner's desire to escape his own family home as well.[13] Hefner recalled that Lownes "felt himself trapped by marriage and green-lawn suburbia":

> He had everything a man could want—a beautiful, loving wife, two fine children, a magnificent home and a good job. The problem was, he was bored beyond belief. He hated tennis club, the endless round of cocktail parties and barbecues, the small talk and the smug respectability of the middle-class American dream. Extra-marital sex, he ruefully reflected, represented his only prospect of excitement. One day in 1953, he simply walked out and never returned.[14]

Soon Lownes was installed in a new bachelor apartment: a single room, ideal for parties, in which the bedroom was located in a curtained recess in one wall. The bachelor pad appeared as a refuge, where the recently-divorced male retreated in search of his lost freedom. This was only the first in a long list of paradoxes involving the construction of an alternative male domestic space: only in the captivity of his apartment could the Playboy be really free. Whereas Steven Cohan, in his article "So Functional for Its Purposes: Rock Hudson's Bachelor Apartment in Pillow Talk," understood *Playboy*'s ideal of bachelorhood as a form "of male liberation from domestic ideology," a cross-reading of *Playboy's* texts and images invites us to interpret the magazine's project as much more paradoxical in relation to domesticity.[15]

Beyond the search for a refuge for the beleaguered male, the Playboy Penthouse articles promoted a position for man within the domestic space traditionally ruled by women. With time, and with the pedagogic aid of *Playboy* magazine, he would learn to reconquer the space that had been expropriated from him by woman, who blinded him with false ideals of marriage and family. Turning Virginia Woolf's plea for "a room of one's own," and the female independence that would come with it, on its ear, the editors of *Playboy* wrote,

> A man yearns for *a quarters of his own*. More than a place to hang his hat, a man dreams of his own domain, a place that is exclusively his.... *Playboy* has designed, planned and decorated, from the floor up, a penthouse apartment for the urban bachelor—a man who enjoys good living, a sophisticated connoisseur of the lively arts, of food and drink, and congenial companions of both sexes.[16]

In the next issue, dedicated to the bedroom and the bathroom, the editors added,

> A man's home is not only his castle, it is or should be, the outward reflection of his inner self—a comfortable, liveable, and yet exciting expression of the person he is and the life he leads. But the overwhelming percentage of homes was furnished by women. What of the bachelor and his need for a place to call his own?[17]

If the 1956 bachelor pad was provided with "the good things that come in leather cases: binoculars, stereo, and reflex cameras, portable radio, and guns," [18] it was not only because the apartment was meant to be a reserve for visual fun and sexual hunting; it was also because it was conceived as a sort of safe and hidden observatory where the male retreats from the dangers of the exterior environment. From the start, the shifting of gender oppositions within American society that brought about the return of the male to the interior space of the home translated into a rejection in the pages of *Playboy* magazine of the political arena, traditionally a male-only territory. "We don't expect to solve any world problems," declared the first *Playboy* editorial, "or prove any great moral truths. If we are able to give the American male a few extra laughs and a little diversion from the anxieties of the Atomic Age, we'll feel we've justified our existence."[19] In the cocoon of his private pad, more or less forgetful of the threat of war but still equipped with the weapons of the last battle, the Playboy could finally freely dedicate his life to the simple joys of consumption and sex.

But was this withdrawal from the political an actual rejection of public space? Was the apparent return to the domestic a sign of the "womanization" of the bachelor?[20] Or was it instead a strategic reaction to the movement of women toward the public arena during and after World War II? What were the limits of the "gender reversibility" of the Playboy?

World War II had radically transformed the feminist debate in America. Whereas the "first wave" of feminists, rallied around equality in voting rights in the 1900s, 1910s, and 1920s, remained faithful to the idea of so-called "separate spheres" (the feminine, naturally tied to maternal and domestic tasks), subsequent waves of feminists were strongly influenced by the outbreak of a war that released women in record numbers into the areas of work and public space. An inverted reaction to the still-unnamed new feminism, *Playboy's* move toward interior space could be read as an attempt to reappropriate traditionally "feminine" space, at exactly the same time when women had gained for the first time in history access to the public and professional realm.

While this movement inward contributed to the active deconstruction of boundaries that had naturalized interior space as feminine and exterior space as masculine and therefore appeared revolutionary, the ideals of *Playboy* were actually supportive of a pre-modern distribution of gendered spaces. Whereas *Playboy's* discourse seemed to be structured as a radical departure from a certain form of old masculinity, its ideal of the "new bachelor," a "city-bred guy—breezy, sophisticated," displayed a rather nostalgic quality.[21] In fact, the magazine's original name, *Stag Party*, referred to a gathering of men assembled for the purposed of watching a "stag film."[22] Produced from the beginning of the twentieth century by men for male-only viewing, these were the first American hard-core pornographic films. In contrast to the color and sound films of the late 1950s and 1960s, watched in movie theatres, these black and white silent films were viewed in private homes—a scenario that stressed male social bonding and camaraderie.[23] The homoerotic structure of the stag party reinforced the notion that not only did men not need women to have fun, but even more, they could have more fun without women. Inspired by these screening parties, Hefner placed his magazine within a particular tradition of "for-men-only" voyeurism. Although first generated through the vehicle of the magazine (which presented nude women for the male gaze), this voyeurism was later encouraged through the screening of Hollywood and pornographic films inside the Playboy House and through the hosting of playmate parties in which men were invited to collectively enjoy "Hefner's girls." By virtually reproducing what could be called "stag space," *Playboy* projected a male retro-paradise into the future.

The ambiguities in relation to both domesticity and gender latent in the definition of the Playboy Penthouse Apartment are manifest in the production of *Playboy*'s Bunny logo, produced as a result of a multi-stage metamorphosis of the "stag." In 1953, a few months before the inaugural issue of *Playboy* was published, Hefner chose a "stag toy"— similar to "Esky," a small plasticine puppet that appeared regularly on the front page of *Esquire* magazine—to represent the magazine. The first drawing of the logo, by Arv Miller, portrayed a male deer wearing a dressing gown and slippers and smoking a pipe. The sketch played with the double meaning of the word "stag" which signifies "the male of red deer" and "a man who attends social gatherings unaccompanied by woman"—conflating the male hunter and hunted deer, outdoor hunting and indoor hunting, into a single image. By transferring Hefner's signature robe and slippers to the animal, it also exhibited an unexpectedly domesticated touch.

Before the name of the magazine was officially established, Hefner learned that "stag" was already used by a field-and-stream magazine. After a brainstorming session, Hefner's friend Arthur Paul suggested the name *Playboy*. Hefner was fascinated by the name, but because he wanted to keep the image of the stag he proposed a slight transformation of Miller's drawing: instead of a deer, the magazine's logo would be a "cute, frisky and sexy rabbit in a tuxedo."[24] By the time Paul finalized the design, the stag had become a "Playboy Bunny"—an unaccompanied, childish male hunting the female sex without leaving home. By January 1954, the male "bunny" had been transformed into a female "bunny."[25]

PENTHOUSE MADE PLAYBOY

Playboy magazine's most urgent mission was to take back the house, because only the interior space, as a gender performative machine, could effectuate the transformation of the man into the Playboy. The text that accompanied the drawings and photographs of furniture in the 1956 article "Playboy Penthouse Apartment: A High Handsome Haven for the Bachelor in Town" presented a double narrative to the male reader. First, the visitor's tour touted the advantages of the "new" management of space within the penthouse, which would allow the bachelor to convert work space into leisure space, a private area into a party hall. Second, the user's guide addressed the reader as a potential consumer of the new space and its functional objects. The underlying seduction tale brought these two narratives together, introducing the middle-class, sexually unsophisticated American male to the management of multiple sexual encounters within a single apartment and presenting sex as the ultimate object of consumption; the management of the interior space amounted to the management of the bachelor's sexual life.

The penthouse's particular value was its ability to produce a gender economy different from that found in the single family home. According to the article, only the "flexibility" of the apartment, the "multiple functionality" of its open space, and the playful, "flip-flop" character of its furniture, embodied in the designs of Charles

Eames, Eero Saarinen, and Osvald Borsani, would render possible the introduction into the house of (as many) women (as is deemed necessary) to satisfy the bachelor's sexual desire. These same features of the penthouse interior would help to protect the bachelor's space from female domestication.

For the first time, the Playboy could be flippant about women, thanks to the apartment's "flip-flop" devices that mechanize flirting. Saarinen's Tulip chairs, a turning cabinet bar, sliding screens, and translucent drapes behave as *dispositifs* of rotation that constantly restructure the space of the apartment to technically assist the bachelor's efforts in defeating the female visitor's resistance to sex. The *Playboy* article maintained,

> Speaking of entertainment, one of the hanging Knoll cabinets beneath the windows holds a built-in bar. This permits the canny bachelor to remain in the room while mixing a cool one for his intended quarry. No chance of missing the proper psychological moment—no chance of leaving her cozily curled up on the couch with her shoes off and returning to find her mind changed, purse in hand, and the young lady ready to go home, damn it.[26]

On one side of the living room, the Saarinen Womb chair could be moved to the right or to the left, transforming a working area into a cruising area (and vice versa) and minimizing the bachelor's waste of time. Moreover, Saarinen's and Eames's intent to design "a comfortable chair, which would allow several sitting positions rather than one rigid one, and [incorporate] a number of loose cushions" fit perfectly within the "work is leisure" agenda of *Playboy*.[27] The "flip-flop couch," raised in the *Playboy* article for its ability to mechanize seduction, was Borsani's Divan D 70.[28] With the D 70, and also the P 40 chaise lounge, Borsani brought into industrial design a rhetoric of mutation, mobility, and flexibility that would become central to *Playboy's* spatial economy. Thanks to a transversal steel mechanism, the divan could be transformed into a bed: "The rest of the living room is best seen by utilizing a unique feature of the couch. It flips, literally: at the touch of a knob at its end, the back becomes sit and vice versa—and now we're facing the other way."[29] No need for convincing the guest; the flip-flop couch converts a casual talk around the table into a romantic *tête-à-tête* in front of the fireplace. This *dispositif* of rotation enabled the bachelor to transform his female visitor, with charm and delicacy, from the vertical to the horizontal position, from woman to bunny, from dressed to nude. With just one more flip-flop movement, the Playboy could take his guest/prey from divan to platform bed—the final trap.

> Now we've slipped the nocturnal dram and it is bedtime; having said "night-night" (or "come along, now, dearest") to the last guest; it's time to sink into the arms of Morpheus (or a more comely substitute). Do we go through the house turning out the lights and locking up? No sir: flopping on the luxurious bed, we have within easy reach the multiple controls of its unique headboard. Here we have the silent mercury switches and a rheostat that control every light in the place and can subtly dim the bedroom to just the romantic level. Here, too, are the switches, which control the circuits for front door and terrace window locks. Beside them are push buttons to draw the continuous, heavy, pure-linen, lined draperies on sail track, which can insure darkness at morn—or no.[30]

It is on this technical platform, closer to a military observatory and control station than a common bed, that the bachelor finally executes his (who knows how technical) sexual performance.

The Playboy Penthouse functions first as an office, or command station, where the bachelor organizes his multiple sexual encounters, and second as a site for those encounters. Once the female guest has entered the apartment, every furniture detail operates as a hidden trap that helps the bachelor to get what *Playboy* magazine calls "instant sex."[31] Mechanical gadgets transform the old ways of hunting the stag into new forms of sexual management proper to the Playboy.[32] The activities of bringing in and eliminating women are enabled precisely by various *dispositifs* of rotation, or flip-flop devices. According to *Playboy* magazine, in addition to assisting with the management of time, these technical accessories prevent two female guests from encountering each other within the space of the apartment and prohibit the "insidious wanting-to-be-a-wife girl" from taking it over altogether. For instance, the phone is equipped with "on-off widgets... so that the jangling bell or, what's worse, a chatty call from the date of the night before won't shatter the spell being woven. (Don't worry about missing out on any fun this way: there's a phone-message-taker hooked to the tape recorder.)"[33]

The effective training given by *Playboy*: first, to get rid of women after sex; second, to eliminate their traces; and third, to prevent women from taking back the kitchen (until now their domestic headquarters), radically transformed the image of the bachelor. The Playboy was no longer a childish or sophisticated lover but rather the technically assisted serial seducer that compulsively responded to the strict demands of a space whose central character was paradoxically at once heterosexual and gender-segregated. In fact, the Playboy, in constant need of eliminating the traces of the night-before date, appeared more like a control-freak, spy, or meticulous serial killer than as a spontaneous, flirtatious bachelor.[34]

THE KITCHENLESS KITCHEN

Articulating sexual difference around the opposition male–technical/woman–natural, *Playboy* magazine maintained that the new domestic environment, saturated with mechanical and electrical appliances, was the rightful domain of the machine-wise male. While the feminine press of the period made efforts to redefine the role of the modern housewife as a technician or manager of the home,[35] *Playboy* would claim that men and not women, trained professionally as tool-makers and machine-operators, were most suited for carrying out newly-automated domestic tasks. The design of the so-called "kitchenless kitchen" in *Playboy*'s Penthouse Apartment, evoked several times by the magazine's editors during the 1960s, signalled this redefinition of a traditionally feminine space as masculine.

The kitchen is camouflaged from the rest of the penthouse—an almost totally open space—by a fiberglass screen. Behind the screen, the interior can hardly be recognized as a kitchen. Every cooking and cleaning appliance has taken the form (at least to the period observer) of a highly sophisticated piece of technology:

> The kitchen walls consist of six Japanese-style Shoji screens, which can slide to completely close or completely open the kitchen. Frames are of elm, covering its translucent fiberglass.... Now let's roll back those Shojis and enter the kitchen. Your first thought might be, where is everything? It's all there, as you shall see, but all is neatly stowed and designed for efficiency with the absolute minimization of fuss and *hausfrau* labor. For this is a bachelor kitchen, remember, and unless you're a very odd-ball bachelor indeed, you like to cook and whomp up short-order specialties to exactly the same degree that you actively dislike dishwashing, marketing and tidying up.[36]

The surprised exclamation of the visitor, "Where is everything?" does not result from the technical character of the appliances, which was a constant in American advertisements for the kitchen at the time.[37] Rather, the word "everything" replaces the word "housewife" in a Freudian slippage. Cleaning, considered by *Playboy* as typical "*hausfrau* manual labor," has been taken over by machines, transforming the kitchen into a playground for the young connoisseur of meat and wine.[38] All of the redefinition of kitchen activities in terms of technical efficiency and male skill safely eliminate any risk of feminizing or emasculating the bachelor. Rejecting at once the "antiseptic medical look of so many modern kitchens" and the feminine character of kitchen appliances, *Playboy* succeeded in making the technical kitchen a necessary accessory, as important a component of the urban stag lifestyle as the automobile.

The technical, kitchenless kitchen takes over the traditional "feminine" tasks of transforming dirty into clean, raw into cooked, not through the efforts of the housewife's working hands but through the aid of machines. The kitchen's ultrasonic dishwasher, which uses inaudible hi-fi sound to clean its contents, eliminates the need for manual dishwashing. The morning after a successful conquest at home, breakfast is prepared by the flick of a remote-controlled switch installed on the bachelor's bed panel. *Playboy* advised the bachelor:

> Reaching lazily to the control panel, you press the buttons for the kitchen circuits and immediately the raw bacon, eggs, bread and ground coffee you did the right things with the night before... start the metamorphosis into crisp bacon, eggs fried just right, and steaming-hot fresh java.[39]

In the kitchenless kitchen, women have been effectively replaced by mechanical appliances that are now under the control of men.

Technical appliances not only come to stand in for the figure of the housewife but also help the serial seducer eliminate all traces of the women who visit the penthouse. Thus, the dishwasher is not only convenient because it is noiseless but also because it removes "the imprints of the lipstick kiss" from the night before.[40] Like the sliding screen of the kitchen, the bachelor's female guests operate under the same visual law: Now you see it, now you don't.

Playboy interpreted the process of transforming the private domestic space of the kitchen into a public showroom—a process generalized within America during the 1950s—as a direct effect of transforming the kitchen into an exclusively male territory. The fallacious logic went that the kitchen was going public because it was becoming a masculine space. With regard to the male user of the "radiant broiler-roaster," *Playboy* wagered, "It is our bet that the manipulation of this broiler, and the sight through the dome of a sizzling steak, will prove for your guest a rival attraction to the best on TV. And you'll be the director of the show."[41] It is as if, for *Playboy*, the transparent dome broiler—like the apartment itself, with its glass windows and undivided spaces—would imitate the structure of the TV set or the show window. These mechanisms of display offered the desired object (the roasted meat, the pink flesh of the Playmate) to the male eye in as fresh and real a manner as possible.

Finally, privacy—meaning total female exclusion—is preserved within two enclosed spaces inside the bachelor apartment: the study, a "*sanctum sanctorum* where women are seldom invited"; and the lavatory, which includes "john, bidet, magazine rack, ash tray and telephone" and which *Playboy* described as a "throne room"—the ultimate retreat, where the bachelor-king "gets away from everything."[42] This urgent privatization of intellect and intestine indicates the limits of the bodily construction of the Playboy: whereas his eyes, hands, and penis are totally devoted to the maximization of sexual pleasure, his reasoning and anality are protected from the menaces of feminine dumbness and homosexuality.

THE INVENTION OF "THE GIRL-NEXT-DOOR"

In the November 1953 editorial of *Playboy* magazine, Hefner affirmed, "We want to make it very clear from the start, we aren't a family magazine. If you're somebody's sister, wife or mother-in-law and picked us up by mistake, please pass us along to the man in your life and get back to the *Ladies Home Companion*."[43] Among middle-class American women, the editorial created a strong reaction. *Playboy*'s anti-family and anti-marriage discourse, together with its presentation of the new bachelor as gender-flexible and as creator of a new type of domesticity, seemed to put into jeopardy the woman's status as wife, mother, and housekeeper. In its January 1959 issue, *Playboy* published a letter of complaint from Mrs. Rose Marie Shelley, of Emporia, Kansas:

> A woman who accepts her husband's celebrating the appeal of other women becomes, in reality, nothing more than his legal bitch; certainly not a real woman or wife—much less a mother worth the title. The nation doesn't need more "understanding women" but more men and women who make their marriage vows on their wedding day and stick to them—without exception. . . . Since when is it man's "prerogative" to practice licentiousness, philandering, adultery, etc.? How can women possibly give men a rank of superiority, when men don't have character or conscience? Your playboys will have to earn women's respect before you ever establish your male supremacy! Show me the woman who doesn't agree![44]

The rhetorical strategy of *Playboy* was to invert the very logic of gender complementarity that ruled the narratives of the American dream, according to which heterosexual and conjugal love was defined by women who take care of the home and men who manage the troubles of the external world. Together, they formed a unit of reproduction and consumption that assured the economic growth of the postwar nation. Against this gendered division of territory, *Playboy* would claim the necessity for men to regain the space of the home. And against the romantic myth of the "loving couple," it would redefine masculine charm in terms of maximizing sexual encounters with women.

The Playboy's sexual success and his spatial conquest depended on the exclusion of three forms of femininity—the mother, the wife, and the housewife—from his new domestic realm. But rather than depicting women as "bitches," as the Kansas housewife had suggested, *Playboy* relied on a spatial strategy to produce man's ideal sexual companion. Placed right at the threshold of the bachelor's own house, accessible and yet separate from his own domestic environment, the "girl-next-door" was to become the new raw material from which to build the ideal Playmate. Years later, Hefner reflected on the creation of this ideal, relating the emergence of the Playmate to the feminist movement:

> The playmate of the month was a political proclamation. *Playboy* wanted to realise an American dream, inspired by the pin-up illustrations and photographs of the thirties and forties: the idea was to transform the next-door neighbour girl into a sex symbol. And this implied a number of changes in relation to the issue of feminine sexuality, meaning that even the nice girls enjoyed sex. It was a very important message, as important as all the feminist disputes.[45]

If the Playboy is the central figure of an ongoing production of an interior and yet not domestic space by men, the Playmate is the anonymous agent of the resexualization of the every-day life. Hefner called this in-your-neighborhood campaign of resexualization "the-girl-next-door effect."[46]

> We supposed it's natural to think of the pulchritudinous Playmates as existing in a world apart. Actually, potential Playmates are all around you: the new secretary at your office, the doe-eyed beauty who sat opposite you at lunch yesterday, the girl who sells you shirts and ties at your favorite store. We found Miss July in our own circulation department.[47]

The Playmate made her debut in the second issue of *Playboy* magazine in 1953, when Jane Pilgrim, Hefner's secretary and lover, appeared as Miss July. Omitting any mention of her liaison with the boss, the magazine presented Pilgrim as a secretary who came to Hefner looking for an Addressograph machine in light of an increasing volume of subscriptions. Hefner said that he agreed to buy her one if she would pose nude.[48] The Pilgrim effect enabled Hefner to identify and improve the girl-next-door formula.

Rather than a natural being to be found right around the corner, as Hefner maintained, the Playmate was the result of a number of representational strategies. The first photographs published by *Playboy* belonged to the American tradition of pin-up paintings. Alberto Vargas, one of the most glamorous pin-up artists of the period, worked exclusively for *Playboy* after 1957. By then, he was known for his aquarelles for the 1927 film *Glorifying the American Girl*, for his movie posters for Twentieth Century Fox, and for his collaboration with the magazines *Esquire* and *Men Only*. The muted colors and airbrushed texture of Vargas's earlier paintings were replaced by the saturated colors and well-defined edges of photographs, which gave the Playmate a quasi-three-dimensional, hyper-real appearance. For *Playboy*'s first issue, Hefner wanted to publish nude photographs of Marilyn Monroe as three-dimensional images to be seen with the aid of glasses, but he abandoned the project due to its high cost. After the success of Monroe's (two-dimensional) spread, the publisher concluded that the contrast of colors and textures—of red velvet against white flesh—produced a similar effect. The Playmate would combine the fleshy and childlike body of the anonymous American pin-up and the glamour of the Hollywood poster girl with the audacity of the pornographic painting.

The *Playboy* Playmate was the pin-up girl come to life. Vargas, as well as other pin-up painters such as Gil Elvgren and Earl MacPherson, mastered the technique of transforming a scene of the "every-day life" of an "American girl" into a carefully staged, color image, ready for mechanical reproduction and distribution. For its part, *Playboy* invested the image with performative power to make "the Varga girl" actually real.[49] The magazine would soon provide its photographers with more than twenty real-life Playmates, who lived night and day inside the pseudo-domestic setting of the Playboy House, from which to choose. In a sense, however, nothing seemed to have changed: the pin-up girl lying on a red modern armchair by Lou Shabner was simply replaced by a Marilyn Monroe look-alike on a red Saarinen armchair.

The realistic rhetoric of *Playboy*'s photographic articles would finally come to show American girls inside out. Elvgren pin-up drawings, in agreement with pornography and X-ray techniques of the 1950s, belong to the same technology of the production of the "true vision" that made possible the exteriorization of interiority and the public production of the private at work at the Playboy Mansion.[50] As the poster by Edward D'Ancona makes manifest, American pornography would ultimately close the circle of production of what Foucault called "the truth of sex,"[51] started by European medical science at the end of the eighteenth century; the Playmate—both pure representation and girl-next-door—is at once "the naked truth," the "bare facts," and "the body."[52]

Within the magazine, the Playmate was the result of the actual visual transformation of the girl-next-door into the naked truth of sex. The unfolding of the four-page centerfold assures the reversibility effect. On the first page, the girl-next-door is represented in what Hefner called "her natural habitat," meaning, for the most part, her house or the office where she works (usually as a secretary).[53] In this image, she performs the role of a rather helpless and infantile girl. The key to this representation of the Playmate, as a possibility already embedded in any girl-next-door, was to distinguish her not only from the "nasty girl" and the common prostitute but also from the predatory woman. The attraction of the Playmate, explains

Russell Miller, "was the absence of threat. Playmates were nice clean girls; there was nothing to fear from seducing them."[54] Unfolding the next, three-page, image, the same girl appears nude, thanks once more to a *dispositif* of rotation similar to the flip-flop couch. The male operation of turning the page transforms folded into opened, hidden into exposed, the next-door-neighbor into the Playmate, dressed into undressed, and finally, "peeping" into "instant sex." And vice versa.

Readers as different as John Berger or Laura Mulvey have observed in their analyses of the European tradition of representing nudes that the real subject of pornographic representation is precisely the male/eye that has been carefully excluded from the picture.[55] However, a sign of the seducer, as a trace of his power to produce the image, completes the frame out of which he has wisely stepped right before the picture is taken. In the case of *Playboy*'s nudes, there is always a sign of the gazing eye in the picture. Every Playmate is represented, no matter how nude, in relation to an exterior object, the most common of which during the 1950s and 1960s was the telephone.[56] The object—a remainder of masculine technology (telephone, hammer, car, etc.) or a clearly identifiable sign of masculine performance (pipe, tie, cigar, etc.)—unpacks the seduction narrative behind the image.[57] There is a seducer behind every Playmate; or, to be more precise, it is the seducer's gaze that transforms the ordinary girl-next-door into a Playmate. At once the seduced eye and constituting eye that enables the transformation of any woman into a real Playmate, the Playboy, at the very limit of representation, haunts the pin-up image.

PIN-UP ARCHITECTURE

How does one interpret the fact that the Playboy Penthouse Apartment articles became the most popular feature published by *Playboy* magazine at the time—as popular as the pin-up photographs of nude Playmates and the articles about their private lives?[58] If the sketches of the Playboy penthouse achieved the same level of notoriety as the nude pictures of Marilyn Monroe, it was because both depended on the same pornographic logic: excitement was directly proportional to the degree of visibility of what was supposed to be private and concealed. Representations of the bachelor pad ignored the structure of the building as well as its exterior appearance, providing exclusively and without exception a view of the interior space. The penthouse articles offered for the public gaze the private scenarios in which multiple and endless seduction encounters could take place. For the first time, the apartment was not used as a mere stage for pornographic pictures, and the presence of nude women was not even required. The Playboy Penthouse Apartment had become *Playboy*'s pornographic object *par excellence*.

In fact, the idea of making public the interior of a bachelor's apartment preceded the publication of the *Playboy* penthouse. Two years before inaugurating *Playboy* magazine, Hefner had tried—unsuccessfully—to convince the *Chicago Daily News* to feature his apartment in a two-page spread under the heading, "How Does a Cartoonist Live?"[59] Once *Playboy* was launched, he pursued his idea through what could be called "surrogate houses": first, with a story about a visit to the fictional Playboy Penthouse Apartment; and later, in May 1959, with a photo-reportage shot inside the bachelor house of Harold Chaskin, Hefner's friend, at Biscayne Bay in Miami.[60]

The success of the Chaskin house piece showed Hefner that nothing was more convincing than using a "real" bachelor pad as the setting for *Playboy's* nude and sexual scenes. Nothing looked as intimate and private as an authentic house. As Gretchen Edgren noted, Chaskin's interior swimming pool "invited women to pose nude," the "subaquatic bar room" (a glass-walled room that enabled the visitor to view the interior of the swimming pool) allowed dressed men to observe the "swimming beauties," and "the master bedroom which overlooked the swimming pool... included an over-dimension bathtub that could easily welcome half a dozen women."[61]

Coming back from Miami, Hefner began developing a plan to build a house that would combine characteristics of the Chaskin house with some inventions of his own.[62] Soon after, he bought a plot of land at 28 East Bellevue in Chicago and commissioned the architect R. Donald Jaye to design a multi-story house around an indoor swimming pool.[63] Somehow, through the influence of either the Catholic city government or the local Mafia, Hefner was denied permission to set his building among the old residences of East Bellevue.[64]

Hefner achieved another hit by publishing the design of the fictional Playboy Penthouse, with interior decoration by Jaye, in May 1962. The drawings showed the glassed facade, furnished section, and some interior details of the bachelor house. The three-story building with strong modern allure literally appeared pasted in between two traditional Chicago houses dating from the turn of the century. Whereas a single brownish color covers both the brick walls and the windows of the adjoining houses, underscoring the opaque character of their facades, the penthouse is made out of a combination of white reinforced concrete walls and wide glass panels. Jutting out from the roof, a small visor holds several lights that illuminate the facade at night, rendering even more visible the spectacle inside. The second floor of the building, housing a living room with a spiral staircase, is totally open to public view. The ground floor is also exposed and shelters a bright blue Porsche. Among the furniture, the piece that created the greatest sensation among *Playboy's* readers was the round, rotating, and shaking bed, equipped, as was the 1956 penthouse, with a control panel, telephone, radio, bar, and night table.[65]

The sectional drawing reveals that the house is symmetrically divided by a large central open space, at the bottom of which is an irregularly shaped swimming pool, or rather a cave, as if the house had been constructed on the very edge of a water source. Although the rooms of the apartment seem rather repetitive, as if multiple and similar scenes could be happening in many places at the same time (the same living room, with its Eames armchairs, is reproduced three times), the sharp split created by the swimming pool between the front and back of the house operates as an exchange passage that modifies the nature of the space. In fact, this division reinforces the duality of the Playboy's life, articulating the transition from work into leisure, dressed into nude, the professional visit into the sexual encounter. Here, the swimming pool functions at once as the *dispositif* of rotation that enables the Playboy to move from the front of the house to the back and as a liquid frontier that separates two separate "stages," where different (and even incongruous) actions take place. The dual structure of the house, as the advertisement for the Porsche suggests, "lets the playboy lead a double life."[66]

THE MAKING OF THE MULTI-MEDIA PLAYBOY HOUSE

By the time the Playboy Penthouse was published in 1962, Hefner was living in his own Playboy house. In December 1959, Hefner bought a regal brick and stone townhouse built in 1899 at 1340 North State Parkway on Chicago's Gold Coast, not far from Lake Michigan. In the early twentieth century, the house had been the center of an intense social scene. During the Great Depression, it had been converted into apartments; nevertheless, the second floor retained the structure of a public house, with its large marble fireplace, ballroom, and hotel kitchen. The 6,600-square-foot house was to be rebuilt by Hefner and represented in the media as an enormous "private" bachelor penthouse.

The cost of the improvements to the Playboy House greatly surpassed its purchase price. The transformation, while if not visible from the outside (Hefner left the exterior facade untouched), affected the very foundations of the house: the six-car garage in the basement was rebuilt as a swimming pool (even though Hefner could not swim) and a "subaquatic room," a human—mostly female—aquarium similar to the one Hefner had seen in the Chaskin House in Miami. In February 1960, Hefner opened a Playboy Club some blocks away from the house; bunnies who worked as escort girls or waitresses at the club lived in the Playboy House. When the renovation was complete, Hefner—as if he himself were the main character of an autobiographical novel—confessed that he only needed a few additional accessories to finally become the Playboy that his magazine had created: a pipe, a smoking jacket, and a Mercedes Benz Cabriolet 3000SL. Finally, a total multimedia architectonic scenario was set and ready for action.[67]

THE INHABITANTS/ACTORS

The inhabitant of the Playboy House is... the Playboy. He is "foot-loose and fiancée free," a young single man or a middle-aged man rejuvenated by the freedom of a recent divorce who enjoys his job as if it were leisure time.[68] He is able to reap the benefits of his party hours since he has perfectly understood that any romance is always business and involves some sort of sportive activity. As Edgren writes,

> He can be a sharp minded young business executive, a worker in the arts, a university professor, an architect, or engineer. He can be many things, provided he possesses a certain *point of view*. He must see life not as a valley of tears, but as a happy time; he must take joy in his work, without regarding it as the end and all of living.[69]

The Playboy is a "man-about-town," an urban bachelor, following the best tradition of European maleness, keeping in touch with his animal side. That is why he is a bullfighter and a hunter.[70] He is a seducer and a sophisticated lover, a man that "could uncouple any known bra with one hand while expertly mixing a martini with the other," leaving his partner topless before she is able to react, while he keeps his smoking jacket intact. He can elegantly practice the "lover's leap" when it is necessary to reach "the distance between twin beds."[71] He is the one and shall remain, without wife or family, one, defying the traps set for him by women. Retreated to his toilet and armed with his TV camera, he is the king. Many would try to equal him, but without success. He is H.H.!

If the Playboy is defined by his singularity, the Playmate is pure multiplicity, ambiguity, and impermanence. For *Playboy*, she is the secretary that takes care of the office and of her boss as if they were her home and her husband. She is a pin-up girl, a nice young woman ready to get undressed in front of a camera. She is a bunny; she likes to wear little bikinis bought in Saint-Tropez, to spend hours in the jacuzzi, and more than anything, she enjoys sex—that is the reason why, according to *Playboy*, even when she is paid for it, "she is not a prostitute," although most of the time she likes to do it for free, or in exchange for fancy European fashion accessories. As Miller explains, "The Bunny is the girl next door. She is the American romanticized myth... beautiful, desirable, and a nice, fun-loving person. A Bunny is not a broad or a 'hippy.' She may be sexy, but it is a fresh healthy sex—not cheap or lewd."[72] The Playboy should be cautious not to become the victim of women who might resemble a Playmate but who could be dangerous to him, like the "gold digger" (a "girl who's got what it takes to take what you've got"), the "under-cover agent," and, especially, the "zombie," the nice unmarried girl looking for a husband."[73] Briefly put, she is any woman, except his wife. A 1960 *Playboy* article concludes, "How then to recognize a Playmate?... Very simple, she is 'instant sex.' You just add Scotch."[74]

ACTION

On the second floor of the Playboy House, the windowless thirty-foot-wide hall with oak-panelled walls adorned with carved frescoes was transformed into a party and screening room. It was here that Hefner held his famous Friday night parties. He also staged his "Playboy's Penthouse" television show, which began in October 1959, here.[75] The broadcast concept for "Playboy's Penthouse" was created by two independent producers in Chicago who approached Hefner with the idea of presenting the Playboy lifestyle on television. The setting would be his bachelor pad, "the kind of paradise every guy is looking for, a place of parties, full of pretty girls and show business celebrities."[76] "Playboy's Penthouse" ran for twenty-six weeks but was never picked up by a national network.

The TV show operated as what Michel Foucault would have called an "inverted mirror,"[77] projecting into the domestic space of television viewers the anti-domestic interior of the Playboy House. The Latin sign on the front door announced, "*Si non oscillas, nili tintinare*," ("If you don't swing, don't knock"). Nobody seemed to be excluded; the only requirement for entrance was to be ready to have fun. The same Chicago that praised the family, embraced Prohibition, and promoted the racial segregation of spaces enjoyed the phantasmatic production of a televisual "deviant heterotopia," ruled by female nudity, polygamy, sexual promiscuity, and seeming racial indifference.[78] In Foucault's terms, the Playboy House functioned effectively as a virtual "counter-site," a sort of "enacted utopia" or "heterotopia" that simultaneously represented, contested, and inverted American sexuality during the late 1950s and 1960s.

As Foucault foresaw, however:

> In general, the heterotopic site is not freely accessible like a public place.... [T]he heterotopias that seem to be pure and simple openings, generally hide curious exclusions. Everyone can enter into these heterotopic sites, but in fact it is only an illusion: we think we enter whereas we are, by the very fact that we enter, excluded.[79]

At the Playboy House, the possibility of visually penetrating the private space was, as in a good Foucauldian heterotopia of deviation, just an illusion; the interior space had been carefully staged and illuminated, like a Hollywood set, and every corner was monitored by a closed-circuit television camera. Thus, when entering the house, the guest, feeling privileged for having been admitted into Hefner's private haven, was actually stepping into virtually public territory. The price paid by every guest for access to this exceptional form of inhabitation was the act of becoming an anonymous actor in an ongoing movie. Again, the same logic of reversibility that ruled the flip-flop couch or the rotating bed transformed the visitor into actor, hidden into displayed, and, of course, private into public. And vice versa.

Within this liminal and heterotopic space, it was necessary to reproduce and reinscribe endlessly new "private" areas seemingly reserved for the happy few but always subject to the surveillance of the camera. A round hole in the floor visually connected the first story of the house to the basement. A golden pole, similar at once to the hyper-masculine fire brigade exit and to a female striptease bar, provided physical access. On one side of the basement was a swimming-pool and cave that imitated the scenery of a tropical island, with palms, bamboo, and a spring running with water; on the other side, separated by a large sliding door, were the garage and the rear exit.[80]

The abysmal character of the hole through which the guest literally "slipped in" and the interior waterfall suggested that the innermost space of the house was its only and real opening.[81] In fact, the cave was the main stage for the photo-reportage that took place in the house and was probably used as well as the setting for pornographic movies.[82] As Louis Marin has pointed out in his reading of Thomas Moro's *Utopia*, it is a characteristic of utopian enclosures to be penetrated right at their center by an empty space, as if the very foundation of the ideal site were precisely a constitutive hole or nurturing non-space. The swimming pool in the Playboy House was like a watery womb that simultaneously produced not only the water but also the women that inhabited the cave, and served as a cosmic drain through which the water and the women flowed away.[83]

A series of pictures of the interior of the house show the double work of the "hole" as opening and discriminating gate. Only the select Playboy could descend to these (technically regulated) natural sites to enjoy a group of nude women waiting just for him. One of the photographs published in *Playboy* magazine shows the less fortunate visitors who remain on the first floor looking through the opening. They appear eager and frightened, as if the very foundations of the house were in danger. Peeping at the Playmates lying in the cave, they seem to be convinced that to fall into that "hole," to penetrate that opening, is the very condition of possibility for masculine sexual enjoyment. In the rear of the basement, in the "subaquatic room," Hefner contemplated the cave party through a window, as if he were watching the next "Playboy's Penthouse" TV show.

THE ROTATING BED

In this endless and reversible production of private/public space, Hefner's rotating bed was the most outstanding apparatus of all. Rejecting the main sleeping configurations of the 1950s, the double bed and two twin beds, Hefner chose a bed larger than the double, yet one that assured the independence and gender-segregation of the twin.[84] Eight and one half feet in diameter, the round bed had an internal motor that allowed it to rotate (although not very smoothly) 360 degrees in either direction and to vibrate while at a standstill.

The leather headboard provided both back support and a control panel from which to operate a radio, television set, film projector, and various telephones. A television camera, mounted on a tripod and pointed directly at the playground of the bed, enabled Hefner to record his "private" business and sexual encounters.

Hefner's exceptional everyday life was the result of a radical inversion of the metaphysical relationship between representation and reality. Thus, the furniture of the Playboy House would mimetically reproduce every detail of the fictional drawings and stories previously published by the magazine. The rotating bed was an improved and hyperbolic version of the rectangular bed that appeared in Joyce's 1959 drawings for the urban Playhouse.[85] As part of a total reciprocity between the private and the public, between the publisher's life and his magazine, Hefner's bed returned to the pages of *Playboy* in April 1965 and soon after became the "most famous bed in America."[86]

In his history of *Playboy*, Miller has suggested that the rotating bed was one of the enigmatic features of Hefner's life:

> Despite heroic efforts, Mr. Hefner was never really able to satisfactorily explain why anyone would want a rotating bed. He used to burble about "creating your different environments" by facing himself in different directions at the press of a button, but it was hard to understand why he could not simply turn his head to achieve the same basic effect. Such was the worldwide media interest in the bed, that a journalist once seriously asked where one bought sheets for an eight-and-a-half-foot diameter round bed. "I haven't the vaguest idea," Hefner answered.[87]

Within Miller's psychologizing argument, the bed was a symptom of "a man who refuses to grown up, who lives in a house full of toys, who devotes much of his energy to playing kids' games, who falls in and out of love like a teenager, and is cross when the gravy is lumpy."[88] In fact, Hefner spent most of his time in bed, always wearing his pyjamas (even in front of guests), eating Butterfingers and candy apples, and drinking Pepsi-Colas.[89] Miller described

Hefner's refusal to leave his bed as pathological, part of a sexual handicap that reduced him to the horizontal position and took him away from the real world, "pampered and cocooned in his citadel of sensualism."[90]

If, as Siegfried Gideon noticed and as Henri Lefebvre regretted, modern social relations are always mediated by objects,[91] this mediation, in the case of Hefner's rotating bed, has been taken to an extreme. As was already the case in the designs for the 1956 Playboy Penthouse and Joyce's unbuilt 1959 Playboy House, in the master bedroom of the Playboy House the machine seems to have replaced the subject's will, prefiguring not only his movements and deeds but also his feelings and decisions. Thus, this living space is neither inhabited nor visited but rather incorporated, the rotating bed serving as a prosthetic Playboy machine into which the bachelor is plugged, as if he were a pre-ambulatory infant or a wounded soldier just back from war. It is this mediating connection that enables him to be in touch with the outside world while remaining profoundly enclosed and that transforms his infantile passivity into sex and business. Rather than understanding Hefner's bed as pathology, it seems more accurate to describe postwar American society as progressively prosthetic, with the rotating bed, at the center of the Playboy House, functioning as its heterotopic site.

THE HORIZONTAL WORKER

In January 1958, *Playboy* published the article "Hollywood Horizontal: Battle Cry of a Vertical Screenwriter. My Kingdom for a Couch." In this essay, the journalist Marion Hardgrove maked public the (fictional) private letters exchanged between William T. Orr, executive producer of Warner Brothers TV in Hollywood, and several Hollywood writers. The ironic quarrel between the voices in favor of and against "verticality" used architectonic criteria to contrast a new type of "horizontal worker," a successful urban writer and businessman, with his "stiff," "vertical" counterpart. Under the command to "take joy in your work," horizontality was understood as the new anti-Weberian ethic of capitalism, whereby work and sex constituted the two main variables in a single equation of male success. One of the writers explains to Orr,

> I have been grievously concerned by recent complaints that my writing is increasingly vertical. In the language of the layman, this means that it goes rigidly down toward the bottom of the page without ever noticeably broadening out. I am stunned by the charge but unhappily unable to refute it. Vertical writing is a serious matter with which we cannot put up. It is a disease that must be treated as soon as it becomes apparent....The bald fact, sir, is this: Horizontal writing cannot be achieved except by being horizontal on a desktop or on cold linoleum. Various secondary officials of our little organization have made conscientious and valiant efforts to stamp out creeping verticality by procuring for the writer that indispensable tool of his trade: the couch.[92]

The floor and not the couch had been the first working surface for Hefner. In his Hyde Park apartment, the undistinguished floor of the different rooms served simultaneously as table, where Hefner spread out his pictures, and as playground. Hefner maintained, " I used the carpet as a giant desk. When I met artists, designers and writers we used to crawl while we looked at our work."[93] The rotating bed of the Playboy House represented a construction above the floor of a second horizontal surface, which Hefner used in an identical manner: he would sit up against the leather back, wearing only his pyjamas, talking on the phone and choosing the next "Playmate of the Month" from among the hundreds of slides. From time to time, and without interrupting his "work," he was visited by a select group of nude bunnies who would become part of his expansive pornographic video archive.

The separation of the home from the workplace was the dominant feature of postwar American urban and suburban life, made possible by the generalized use of the automobile.[94] Hefner's rotating bed—used as work table, television stage, TV couch, sexual playground, orgy site, and sleeping platform—levelled a direct attack against this effective segregation of spaces, against this distancing of the places for work and from the places for recreation, against the rupture between professional and private environments. The rotating bed was the master *dispositif* of rotation, transforming vertical into horizontal, up into down, right into left, adult into child, one into many, dressed into nude, work into leisure, and private into public. And vice versa.

ROTATING THE PLAYMATE, OR HOW TO TRANSFORM VIKKI DUGAN INTO "THE BACK"

The Playboy's mechanical gadgets were not the only things to operate as rotating devices. The same *dispositif* of rotation that would enable Hefner's bed to turn 360 degrees was behind the production of one of the most famous Playmates of the 1950s, known as "The Back." In June 1957, *Playboy* published photographs taken by Sam Baker of Vikki Dugan's nude back.[95] One month later *Playboy* dedicated a three-page story to the new Playmate sensation. The article read:

> At the Hollywood Foreign Press Association's 1957 award banquet, Vikki turned up in a gown that was not only backless but virtually seatless too—cut down to reveal several startling inches of reverse cleavage. Masculine eyeballs popped, as did the flashbulbs of the United Press, who caught Vikki with her rearguard down and sent the wires a fascinating photo that has to be judiciously cropped for newspaper publication.[96]

In the article Dugan denounced the hypocrisy of what she called "people in glass dresses," a criticism of the models who posed in transparent tissue (the most common way of showing a female nude in the classic pin-ups of the late 1940s and 1950s such as those by Vargas). She argued for a different way of showing and concealing the female body. She was portrayed wearing an opaque fabric dress that wildly exposed (to the male gaze) what *Playboy* had judged her sexiest body part: her backside. Once the "best parts" of Dugan were selected, photographed, and cropped, the metonymic process could take place: Dugan became "The Back."

The possibility of "looking things from behind" was not only a consolation for women such as Vikki Dugan who "were not bustly":[97] turning the bustless girl to discover the back of a Playmate was another rotation game through which *Playboy* inverted the laws of the gaze. What was back became front, exactly in the same way that, through the use of the TV camera, the "private" rooms Hefner's house became public and what was hidden became exposed. Like the cropping of Dugan's back, the visibility of the Playboy House was regulated through a very precise selection of images, staged for the public eye. In fact, Hefner used his television show as a way of "focusing in" and "opening" to the public eye some of the staged scenes already published in the magazine, offering what he called (in a phrase that underscored the production of the "private") a "behind-the-scenes view of America's most sophisticated magazine."[98] The Playboy House multi-media *dispositif* assured the interplay between the house, the magazine, and the TV show.

Like Hefner's bed, *Playboy* magazine itself can be understood as a horizontal plane, an ideal grid upon which all of the fragmented body parts of the *Playboy* empire relate to each other, as in a Saussurean structural system. The bed was used literally as a board upon which Hefner played with the pieces of the different pictures. It was within this plane that a particular cropped organ referred to another, by homology or by difference: not only did Dugan's back establish a "flip-flop relationship" with the bust of another prominent Playmate June Wilkinson, but the blonde hair and smiling face of the as-yet-unknown girl-next-door Stella Stevens were analogically linked to those of Playmates Marilyn Monroe and Kim Novak. The two-dimensional space of the photograph, which provides the possibility of cutting and combining different body parts endlessly, was simultaneously the origin and the result of *Playboy's* visual, pornographic classification of women. This space extends itself without relief toward the past and the future equally, embracing every woman that ever existed or will ever exist. It is within this plane of analogies that the girl-next-door, innocent as she might be, is already connected, even without knowing it, to Brigitte Bardot.

Moreover, within this visual chessboard, Bardot herself becomes merely a gracious combinatory formula of Gina Lollobrigida, Jayne Mansfield, Anita Ekberg, and even the forthcoming Vanessa Paradis.

As the pairing of "The Back" and "The Bust" shows, the *dispositif* of rotation establishes a relationship between two objects or body parts that do not necessarily belong to the same owner, exactly the same way that the pornographic montage cuts hands, mouths, and genitals from different sources and pastes them together as part of a sexual narrative. The transformation of Dugan into "The Back" exemplifies a strategy of multiple composition out of which not only the Playmates but also their position in the Playboy House are constructed.

Just as, according to Foucault, "the heterotopia is capable of juxtaposing in a single real place several spaces, several sites that are in themselves incompatible," the Playboy House brought together, through vertical and horizontal distribution, the bachelor penthouse, the TV show stage, the nice little girls' boarding school, and the brothel.[99] The casual appearance of "private" parties full of girls, the "home-like" images of Hefner in the water cave, or the *tableau vivant* of women around the fireplace playing indoor games depended solely on the existence of a well-programmed space called the "Bunny Dorm." Located on the third floor of the Playboy House, right above Hefner's haven, the dorm aimed to deliver, with mathematical precision, a certain number of well-disciplined bunnies to the floors below and, later, to the Playboy clubs.

Whereas the basement, first, and second floors of the house were characterized by glamorous furniture, technical club-like accessories (screening devices, stereo system, etc.), and large, undivided spaces for playing, dancing, and swimming, the third and fourth floors were rarely opened to male visitors. The door to the third floor represented the most radical gender segregation line and border of "privacy" within the building. The floor was divided into several suites, all identified by the color of their décor (blue, red, golden, etc), where Hefner's female friends and associates could spend some time, and several single apartments rented to Hefner's favorite bunnies. The fourth floor was occupied by large dormitories, rooms, communal showers and toilets, long corridors with public phones, and little mailboxes organized by the bunnies' names. As Miller notes, "In stark contrast to the push-bottom extravagance below, the furnishing of the dormitories abruptly takes on the aspect of a rather parsimonious girls' boarding school—thin cord carpet, bunk beds, wooden lockers, and communal washrooms."[100] The backstage and storeroom of the Playboy House was a bunny boarding school, where the girl-next-door was trained to be a Playmate.

Upstairs, a strict, almost military, disciplinary regime replaced the relaxed atmosphere of Hefner's own quarters. Every bunny was recruited after a rigorous process of selection organized by Keith Hefner, Hugh's brother. Upon being hired, the girl was required to sign a contract agreeing to keep her physical appearance and personal conduct "beyond reproach" and, of course, to always be available inside the Playboy House. Once in the dorm, a bunny would pay $50 in rent per month and could eat in the bunny dining room for $1.50, so that leaving the house was rendered unnecessary, if not impossible.[101]

The bunnies were paid $50 a day for posing, "acting," or working at the club; the rest of their income came from tips and clients' gifts. What looked like "a good salary" for a girl-next-door coming from the Midwestern countryside represented less than .05 percent of the profits that she produced for Hefner's business. The profitability of the Playboy House, tentacularly self-reproducing through the media vehicles of the magazine, the TV show, and the Playboy clubs, surpassed that of Chicago's famous brothels.

A ROOM FOR VICE

The author of the article "No Room for Vice," published in the January 1959 issue of *Playboy*, established a close relationship between architecture and sexuality, suggesting that the modernization of America during the postwar years had successfully led to the replacement of the old-fashioned "red light district" and "old theaters of vice" with the new "bachelor quarter." In a similar manner, he opposed the old forms of "prostitution" to a new form of "feminine sexual freedom":

> There aren't any prostitutes in Chicago for the same reason that there aren't any straw hats at the North Pole. They would starve to death.... Every fourth female over 18 in the city of Chicago is very active sexually, either on a romantic basis, or on a financial one. Usually on both.... In addition there must be at least 100 thousand girls living in the bachelor quarters where they are able to entertain their bosses and business associates. I have not... heard of any male Chicagoan complaining about sex frustration. To the contrary.[102]

What is praised in the article is the liberalization of the sexual market; the sexual services previously provided by a small group of women who were considered prostitutes have been "democratized," extended to the ensemble of the American female population. *Playboy* magazine's promotion of the transformation of work into leisure as the main lifestyle guideline for the new bachelor was coupled by the ability of the Playmate to transform sexual labor into entertainment.

Together with this "liberalization," *Playboy* welcomed the sexualization of everyday places, in contrast to the concentration of the sexual market inside the brothel:

> The swanky cafés, full of ritzy-looking callgirls, the hotel suites reserved for out-of-town buyers of *amour*; the back seat of the automobile, the untidy bachelor girl apartments, the dimly lit booths in cheap eateries, the friend's apartment borrowed for the evening with the phone ringing an unnerving obligation—such and countless other improvised *rendezvous* have displaced the old plush and crystal brothels.[103]

But weren't those precisely the same everyday, banal settings chosen as staged sites for *Playboy*'s stories? Paradoxically, the magazine's discourse fought with equal force familiar domesticity and the traditional brothel. As a replacement for both, Hefner invented the perfect sexual heterotopia, an exceptional folding-in of the outside world into a public house, a private brothel, and a virtual form of sexual enjoyment without sex, all under one roof.

The Playboy House should be placed inside the genealogy of brothels, instead of being considered as a special and monumental example of the modern bachelor house. The distribution of public and private spaces inside the house, and its curious conflation of work space and domicile, were in fact not very different from nineteenth- and early twentieth-century brothels.[104] Hefner's success, however, was to convert the early form of sexual consumption, which took place inside the brothel, into pure representation and visual consumption, multiplying the "value" of every one of "his" Playmate's sexual gestures and acts. The exceptional status of the Playboy House was not the result of its being America's biggest bachelor's apartment, as the press would put it, but rather, of its being the first mass-media brothel. If pornography can be understood as any representation of sexuality, the aim of which is the management of the sexual response of the observer, then the Playboy House was nothing other than a multi-media pornographic device.[105]

In terms of its vertical distribution, the house's stairs organized the transition between the restricted space of the Bunny Dorm upstairs, where visitors were not allowed, and the sexual freedom of the floors below, where bunnies were required to always be available to be photographed and filmed. In terms of media production and distribution, the house, with its themed spaces (the tropical cave, the colored suites, the living rooms), produced the flow of images that appeared in *Playboy* magazine and on the "Playboy's Penthouse" TV show.[106] Whereas the images of the interior of the house seemed to convey the intimacy of Hefner's private sanctuary, every photograph was the result of careful staging. What was rendered public was a particular representation of the interior as "private." This process of "the public" construction of the "private" reached its climax with the creation of the Playboy Club—an imitation of the interior of the house outside the house.

THE PLAYBOY CLUB

In February 1960, Hefner opened the first Playboy Club at 116 East Walton Street in Chicago, just a few blocks away from the Playboy House. The club was built as a "public" reproduction of the interior of Hefner's Playboy House, the design following the distribution of space on its first and second floors. "Each of the four floors was designed as a 'room' in the mythical and fabulous bachelor pad—there was a Playroom, a Penthouse, a Library, and a Living Room."[107] The ticket to enter the club was a bunny logo key, similar to the one that appeared in the 1956 Playboy Penthouse article, purchased by visitors for $5. Ruled by the same laws as the Playboy televisual brothel, visitors could look but never touch the more than thirty bunnies that served each floor of the club. Only privileged clients, considered "special guests" rather than mere visitors, were given a "Number 1 Key" authorizing them to entertain and touch bunnies inside certain rooms. The club—as a reproduction of the Playboy House—operated as a surrogate domicile, a sort of anti-domestic theme park, where the anonymous client paid to be able to perform the role of the ideal bachelor for a few hours. By 1963, there were Playboy clubs in New York, Miami, New Orleans, Saint Louis, Los Angeles, and Baltimore. The conquest of interior space, promoted by *Playboy* magazine beginning in 1953, was indeed taking hold.

NOTES

COLD WAR HOTHOUSES

1. Parts of this long-term research project, which is forthcoming as a book entitled *Domesticity at War*, have been published as "Domesticity at War," in *Assemblage: A Critical Journal of Architecture and Design Culture* 16 (1991): 14–41; "DDU at MoMA," in *ANY* 17 (1997): 48–53; "1949," in *Autonomy and Ideology: Positioning an Avant-Garde in America: 1923–49*, ed. Robert E. Somol (New York: Monacelli Press, 1997), 301–25, 355–57; "Reflections on the Eames House," in *The Work of Charles and Ray Eames: A Legacy of Invention* (New York: Harry N. Abrams, 1997), 126–49; "The Lawn at War: 1940–59," in *The American Lawn: Surface of Everyday Life*, ed. Georges Teyssot (New York: Princeton Architectural Press, 1999), 134–53; and "Enclosed by Images: The Eameses' Multimedia Architecture," in *Grey Room: Architecture, Art, Media, Politics* 2 (Winter 2001): 5–29.
2. Beatriz Preciado came up with the title "Cold War/Hot Houses" during the first seminar, which was dedicated exclusively to the impact of the Cold War on domestic architecture. As we moved to larger-scale investigations during the second year, the title was questioned and then kept in the sense described in the text, on page 12. Other possible interpretations included Brendan Hookway's idea that each one of the topics here—from toys to highways—was itself a hothouse during the Cold War.
3. David Snyder, "At the Drive-In," paper presented in the "Cold War Hot Houses" seminar, Fall 2001. Other papers from the seminar not appearing in this volume include: Michael Hermann, "In the Kitchen," Emmanuel Petit, "Plan B," and Ingeborg Rocker, "TV."
4. Jack Kerouac, *The Dharma Bums* (1958; reprinted New York: Penguin Books, 1976), 39, 104.
5. "The push buttons designed to make housework easier came from the same laboratories as the push buttons for guided missiles.... Chrysler, General Electric, Goodyear, and Westinghouse were all major Pentagon contractors." Stephen J. Whitfield, *The Culture of the Cold War* (Baltimore and London: The Johns Hopkins University Press, 1991), 74.
6. "To us, diversity, the right to choose, is the most important thing.... We don't have one decision made at the top by one government official.... We have many different manufacturers and many different kinds of washing machines so that the housewife has a choice." Quoted by Elaine Tyler May in *Homeward Bound: American Families in the Cold War Era* (New York: Basic Books, 1988), 17. For transcripts of the debate, see "The Two Worlds: A Day-Long Debate," *New York Times*, 25 July 1959; "When Nixon Took on Khrushchev," a report of the meeting, and the text of Nixon's address at the opening of the American National Exhibition in Moscow on 24 July 1959, printed in "Setting Russia Straight on Facts about U.S.," *U.S. News and World Report*, 3 August 1959, 36–39, 70–72; "Encounter," *Newsweek*, 3 August 1959, 15–19; and "Better to See Once," *Time*, 3 August 1959, 12–14.
7. See Beatriz Preciado, "Pornotopia," 208 of this volume.
8. While many sources insist, following the Eames, that the new version used only those parts already delivered to the site, with the exception of one additional beam, Marilyn and John Neuhart question this: "A count of the seventeen-foot vertical girders needed for both house and studio yields a total of twenty-two for the first and sixteen for the latter, considerably more that would

have been needed for the first version of each. In addition, there do not appear to have been any seventeen-foot girders in the original house. Additional trusses would also have been required to accommodate the reworked plan." *Eames House* (Berlin: Ernst & Sohn, 1994), 38. See also Beatriz Colomina, "Reflections on the Eames House."

9. Alison and Peter Smithson, *Changing the Art of Inhabitation* (London: Artemis, 1994), 101.
10. "Case Study Houses 8 and 9 by Charles Eames and Eero Saarinen, Architects," *Arts & Architecture*, December 1945, 43.
11. Charles Eames, quoted in John Neuhart, Marilyn Neuhart, and Ray Eames, *Eames Design: The Work of the Office of Charles and Ray Eames* (New York: Harry N. Abrams, 1989), 157.

COCKPIT

1. Today, in U.S. military planning, "transformational" has become the catch phrase for all attempts to rethink the future of the military in the wake of the Cold War. The Armed Forces have increasingly looked toward paradigm shifts, or historical turning points, in the conduct of warfare as a means of thinking through the problem of change, whether in regard to technology, organization, or the global context. Although different theoretical schools of reform exist (under such headings as "Revolution in Military Affairs," "Fourth-Generation Warfare," and Secretary of Defense Donald Rumsfeld's own "Force Transformation"), the common theme is that institutions must view continual change, both external and internal, as a permanent state of affairs. See, for example, Bill Keller, "The Fighting Next Time," *The New York Times Magazine*, 10 March 2002.
2. This list compiled by Mills is revealing: General Lucius D. Clay, who commanded troops in Germany, then entered the political realm as occupation commander, is now the board chairman of the Continental Can Company. General James H. Doolittle, head of the 8th Air Force shortly before Japan's surrender, is now vice president of Shell Oil. General Omar N. Bradley, who commanded the 12th Army group before Berlin, went on to a high staff position and then became the board chairman of Bulova Research Laboratories; in February 1955, Chairman Bradley allowed his name to be used—"General of the Army Omar N. Bradley"—on a full-page advertisement in support, on grounds of military necessity, of the new tariff imposed on Swiss watch movements. General Douglas MacArthur, political general in Korea and Japan, is now chairman of the board at Remington Rand, Inc. General Albert C. Wedemeyer, commander of U.S. forces in the China theater, is now a vice president of AVCO Corporation. Admiral Ben Moreell is now chairman of Jones and Laughlin Steel Corp. General Jacob Evers is now technical advisor to Fairchild Aircraft Corp. General Ira Eaker is vice president of Hughes Tool Co. General Brehon Somervell, once in charge of Army procurement, became, before his death in 1955, chairman and president of Koppers Co. Admiral Alan G. Kirk, after serving as ambassador to Russia, became chairman of the board and chief executive officer of Mercast Inc., which specializes in high-precision metallurgy. General Leslie R. Groves, head of the Manhattan Project, is now a vice president of Remington Rand in charge of advanced research. General E. R. Quesada, of the H-Bomb test, is a vice president of Lockheed Aircraft Corporation. General Walter Bedell Smith is now vice chairman of American Machine and Foundry Company's board of directors. Army Chief of Staff General Matthew B. Ridgway, having apparently turned down the command of Kaiser's automotive invasion of Argentina, became chairman of the board of the

Mellon Institute of Industrial Research. C. Wright Mills, *The Power Elite* (London: Oxford University Press, 1956), 180–81, 214.

3. Vannevar Bush, "As We Might Think," *The Atlantic Monthly* 176, no. 1 (July 1945): 101–8. A reprint of this article, along with a selection of articles detailing the rise of the "military-industrial complex," may be found in Branden Hookway and Sanford Kwinter, eds., "Dossier," in Rem Koolhaas, Stefano Boeri, and Sanford Kwinter, *Mutations* (Barcelona: Actar, 2000), 628–49.
4. See Kurt Lewin, "Frontiers in Group Dynamics," *Human Relations* vol. 1, 5. Lewin helped found the influential journal *Human Relations*, which was published as a joint project by his Research Center for Group Dynamics (now at the University of Michigan) and the Tavistock Institute, London.
5. Ibid., 5.
6. Robert M. Yerkes, ed. "Psychological Examining in the United States Army," *Memoirs of the National Academy of the Sciences*, vol. 15 (Washington D.C.: Government Printing Office, 1921), 91.
7. Robert M. Yerkes, "Report of the Psychology Committee of the National Research Council," *Psychological Review* 26 (1919): 83–149.
8. From a lecture delivered by Major Robert M. Yerkes in New York, N.Y., 25 January 1919, cited in Clarence S. Yoakum and Robert M. Yerkes, *Army Mental Tests* (New York: Henry Holt and Company, 1920), vii.
9. Edward. L. Thorndike, "Scientific Personnel Work in the Army," *Science* 49 (1919): 54. See also Thorndike's seminal work in behavioral science, *Animal Intelligence: An Experimental Study of the Associative Processes in Animals* (New York: The Macmillan Company, 1898).
10. Thorndike, "Scientific Personnel Work," 56.
11. Ibid.
12. Clarence S. Yoakum, "Plan for a Personnel Bureau in Educational Institutions," *School and Society*, 10 May 1919, reprinted in Yoakum and Yerkes, *Army Mental Tests*, 178.
13. See Alfred Binet, "New Methods for the Diagnosis of the Intellectual Level of Subnormals" (1905), in *The Development of Intelligence in Children: The Binet-Simon Scale*, trans. Elizabeth S. Kite (Baltimore: Williams & Wilkins Co., 1916).
14. The Stanford psychologist Lewis M. Terman's revision of the Binet-Simon scale (referred to as the Stanford-Binet scale) was widely used in the U.S. and was based upon tests of 396 Californian children between 1910 and 1911. A total of 2,300 subjects were tested using this scale by 1916. Lewis M. Terman, *The Measurement of Intelligence* (Boston: Houghton Mifflin, 1916), 36–37, 51–54. See also Lewis M. Terman et al., *The Stanford Revision and Extension of the Binet-Simon Scale for Measuring Intelligence* (Baltimore: Warwick & York, Inc., 1917), 7–31.
15. Yoakum and Yerkes, *Army Mental Tests*, 12.
16. Ibid., 18.
17. Ibid., 19.
18. Ibid., xii.
19. Thorndike, "Scientific Personnel Work," 55.
20. Lewis M. Terman, memorandum to the Student's Army Training Corps, delivered several days after the signing of the armistice, cited in Yoakum and Yerkes, *Army Mental Tests*, 160.
21. Yoakum and Yerkes, *Army Mental Tests*, 161–83.
22. In 1922 Terman became president of the National Education Association and pursued a long-term study of gifted children in California, which continued until his death in 1956. A discussion by Thorndike on the aims and results of 1920s intelligence tests, including the Stanford-Binet, the Army Alpha, and the National Intelligence tests, may be found in Edward L. Thorndike et al., *The Measurement of Intelligence* (New York:

Teacher's College, Columbia University, 1927), 469–90.

23. Terman, *The Measurement of Intelligence*, 21.
24. Walter Lippmann, "The Future of the Tests," *New Republic* 32, no. 417 (29 November 1922): 9–11. This was the last article in the magazine's series, which also included "The Mental Age of Americans," vol. 32, no. 412, 213–15; "The Mystery of the 'A' Men," vol. 32, no. 413, 246–48; "The Reliability of Intelligence Tests," vol. 32, no. 414, 275–77; "The Abuse of the Tests," vol. 32, no. 415, 297–98; and "Tests of Hereditary Intelligence," vol. 32, no. 416, 328–30.
25. Lippmann, "The Reliability of Intelligence Tests," 277.
26. Thorndike, for example, faulted the existing Army tests, maintaining "over-weighting ability to think with words and symbols in comparison with ability to think with materials and mechanisms, our whole procedure in measuring intelligence requires a critical review; and probably the common view of intelligence requires reconstruction." Thorndike, "Scientific Personnel Work," 55. Yet this critique is mainly concerned with improving the effectiveness of the tests. Even for Yerkes, "The concrete significance of general testing is difficult to describe.... Correspondence between school success and the tests is relatively high.... Clerical workers succeed in general in proportion to score; but many other factors are to be considered even in these cases of positive correlation.... The most dangerous thing that can happen is to have education, economics, sociology and industry accept the results of mental tests uncritically and with haste for immediate service that does not permit careful study and research." Yoakum and Yerkes, *Army Mental Tests*, 201–03.
27. At the time, some psychologists, including Binet, saw intelligence as a genetic predisposition modulated by environmental factors, while others, such as Terman, saw intelligence as something innate, attributable to heredity and race. This latter view fed into common fears of "racial decline" and "the menace of the feeble-minded" as well as the "science" of eugenics.
28. Yoakum and Yerkes, *Army Mental Tests*, 180–81.
29. From a speech by Raymond Dodge. cited in Yoakum and Yerkes, *Army Mental Tests*, 204.
30. Thorndike, "Scientific Personnel Work," 57–58.
31. Yerkes, "Report of the Psychology Committee," 26.
32. Knight Dunlap, "Psychological Research in Aviation," *Science* 49 (1919): 94–97.
33. Ibid.
34. For more on human factors in military aviation between the wars, see C. G. Sweeting, *Combat Flying Equipment: U.S. Army Aviators' Personal Equipment, 1917–1945* (Washington, D.C.: Smithsonian Institution Press, 1989).
35. Giulio Douhet, *Command of the Air*, trans. Dino Ferrari (Washington D.C.: Office of Air Force History, 1983).
36. Arthur J. Hughes, *History of Air Navigation* (London: George Allen & Unwin Ltd., 1946), 33.
37. Michael Russell Rip and James M. Hasik, *The Precision Revolution: GPS and the Future of Aerial Warfare* (Annapolis: Naval Institute Press, 2002), 16–17. Pilotage was often called "flying the iron compass," after the practice of following rail lines to a target destination.
38. Hughes, *History of Air Navigation*, 33–37.
39. Norris B. Harbold, *The Log of Air Navigation* (San Antonio: The Naylor Company, 1970), 46.
40. Ibid., 10.
41. William C. Ocker and Carl J. Crane, *Blind Flight in Theory and Practice* (San Antonio: The Naylor Company, 1932), 10–11.
42. John A. Macready, "The Non-Stop Flight across America," *National Geographic*, July 1924, 64, cited in Ocker and Crane, *Blind Flight*, 11.

43. Harbold, *Log of Air Navigation*, 47. Sperry also developed gyroscopic direction indicators, gun battery control systems, automatic pilot systems, and stabilization systems for the U.S. Navy during the war. His company became one of the cornerstones of the defense industry, working closely with the Army Air Corps on the gradual invention of instrument flight.
44. Harbold, *Log of Air Navigation*, 47–50.
45. Ocker and Crane, *Blind Flight*, 13–14.
46. David A. Myers, "The Medical Contribution to the Development of Blind Flying," *Army Medical Bulletin* 36 (July 1936), cited in Harbold, *Log of Air Navigation*, 48.
47. Harbold, *Log of Air Navigation*, 48.
48. The three systems of blind flight were the ABC-System, the XYZ-System, and the Artificial Horizon System. In the ABC-System, one would "A. Use rudder to center turn indicator. B. Use Ailerons to center ball bank indicator. C. Use Elevator to center climb indicator." The XYZ used the same instruments: "X—Center the turn indicator using *ailerons* and *rudder* (Coordination of controls), then, Y—*Keep* the turn indicator centered with rudder; center ball inclinometer with ailerons, and, Z—Center climb indicator with elevators; check air speed." The Artificial Horizon System made use of the direction gyro to maintain direction and the attitude indicator for spatial orientation. Ocker and Crane, *Blind Flight*, 121, 128, 144–47. See also Harbold, *Log of Air Navigation*, 51.
49. Ibid., 53–54.
50. From a speech by Brig. Gen. David N. W. Grant, in 1948, cited in Morris Fishbein, *Doctors at War* (New York: Dutton, 1945), 278–79, subsequently cited in Sweeting, *Combat Flying Equipment*, 15.
51. Arthur W. Melton, "Military Psychology in the United States of America," *American Psychologist* 12 (1957): 741.
52. David Meister, *The History of Human Factors and Ergonomics* (Mahwah, N.J.: Lawrence Erlbaum Assoc., 1999), 173.
53. Important work here includes that done at the Aero Medical Laboratory at Wright Field, under the direction of Paul Fitts, and the Applied Psychology Unit of Cambridge University, under the direction of Frederic Bartlett.
54. W. F. Grether, "Engineering Psychology in the United States," *American Psychologist* 23 (1968): 743–51.
55. In addition to the continued work after World War II of the AAF Aero Medical Laboratory, the Navy set up the Naval Research Laboratory in Anacostia, Washington, D.C., the Navy Special Devices Center in Port Washington, New York, and the Navy Electronics Laboratory in San Diego, California. These groups worked in conjunction with engineering researchers at institutions such as the University of California at Berkeley, Harvard, Johns Hopkins, and the University of Maryland. Meister, *History of Human Factors*, 177.
56. Lewin, "Frontiers [in] Group Dynamics," 40.
57. Frederick Winslow Taylor, *The Principles of Scientific Management* (New York: W. W. Norton and Co., 1967), 85.
58. Robert Probst, *The Office: A Facility Based on Change* (Elmhurst, Ill.: The Business Press, 1967).
59. Ibid., 12.
60. Ibid., 14.
61. See Douglas McGregor, "Theory X: The Traditional View of Direction and Control," and "Theory Y: The Integration of individual and Organizational goals," in *The Human Side of Enterprise* (New York: McGraw-Hill Book Company, 1960), 33–57, cited in Probst, *The Office*, 18. Theory Y was first presented in a speech at MIT's Alfred P. Sloan School of Management in 1957.
62. Ibid., 48.
63. Psychologist Abraham Maslow, a colleague of E. L. Thorndike at Columbia University during the 1930s, sought in publications such as

Motivation and Personality (New York: Harper, 1954) and *Toward a Psychology of Being* (Princeton: Van Nostrand, 1962) as well as in the *Journal of Humanistic Psychology* (which he co-founded) to humanize the mechanistic worldview of American behaviorist psychology by incorporating ideas drawn from Gestalt psychology and European existentialism. His famous "hierarchy of needs" proposed a fundamental ranking of human goals, with bare survival at the base and "self-actualization" as the ultimate goal of human existence.

64. "With a gaming technique, it is possible and practical for managers to experiment with variations on the communications structure. Using a mathematical model and a computer he can get immediate response to a 'what if' speculation." Probst, *The Office*, 39.
65. "Third generation computers using Cathode Ray Tube displays with a light pen for input and drum plotters for hard copy read-out provides the large organization planners with a superb tool.... For the large facility, physical scale models become too large and cumbersome. Computer graphics simulation will emerge more and more as the large facility planner's tool." Ibid., 40.
66. Franklin V. Taylor, "Psychology and the Design of Machines," *The American Psychologist* 12 (1957): 249–58.
67. Ibid., 255, 250.
68. Ibid., 254.
69. See Lawrence J. Fogel, "A New Concept: The Kinalog Display System," *Human Factors* 1, no. 2 (April 1959): 30–37.
70. Ibid., 30.

FORECAST

This article was written after 9/11/01, during the invasion and occupation of Afghanistan, and was completed during the commencement of the war in Iraq. This text is dedicated to ideas forgotten, dismissed, or ignored by the RAND Corporation—enlightened diplomacy and a peaceful future.

I would like to thank the following people at the Senator John Heinz Historical Society of Western Pennsylvania, Pittsburgh, Penn., whose assistance with the Alcoa archive material was invaluable: Steve Doell, director of archives, Rebekah Johnston, coordinator of photographic services, and W. Douglas McCombs, assistant curator.

1. Charles C. Carr, *Alcoa: An American Enterprise* (New York: Rinehart & Co., 1952), 257.
2. Lawrence Paterson, *The First U-Boat Flotilla* (Annapolis, Md.: Naval Institute Press, 2002), 56. See also Carr, *Alcoa*, 256. A second group of saboteurs involved in Operation Pastorius were dropped off along the shore in Jacksonville, Florida. Both groups were instructed in the use of explosives with the aim of destroying railroad lines delivering aluminum to aircraft manufacturing plants. The namesake of the operation, Francis Daniel Pastorius (1651–1716), was the founder of the first German settlement in the U.S. during the seventeenth century—Germantown, Pennsylvania.
3. United States Department of Commerce, *Materials Survey—Aluminum* (Washington: U.S. Government Printing Office, 1956), II–9.
4. One of the early projects to implement systems analysis took place in 1948, when the corporation completed an offensive bomber study, recommending for future strategic bombers the use of a turbo-prop airplane. Dissatisfied with the design modifications proposed by RAND, a ranking Air Force officer bluntly remarked, "I wouldn't strap my ass to that RAND bomber under any circumstances." More successful recommendations would follow in an air defense study completed in 1951. Bruce L. R. Smith, *The RAND Corporation: Case Study of a Nonprofit*

Advisory Corporation (Cambridge, Mass.: Harvard University Press, 1966), 104–05.

5. Reinhold Martin makes a similar connection between Cold War organization and domestic periodicals. He writes, "Through the fissures of a bipolar cold war there thus emerged a logic of control so encompassing that it aspired to the status of both material *and* discursive regulator, an organizational 'pattern' encoded in images circulating through the same mass-media networks... that [Marshall] McLuhan analyzed in *The Mechanical Bride*." Reinhold Martin, *The Organizational Complex: Architecture, Media, and Corporate Space* (Cambridge, Mass.: MIT Press, 2003), 37. In his book *The Mechanical Bride*, within the chapter entitled "The Ballet Luce," McLuhan commented on the most influential magazine of the period: "*Time*... is an important factor in contemporary society. Its shape and technique constitute a most influential set of attitudes which are effective precisely because they are not obviously attached to any explicit doctrines or opinions. Like the clever ads, they do not argue with their readers. They wallop the subconscious instead." Marshall McLuhan, *The Mechanical Bride: Folklore of Industrial Man* (Toronto: Copp Clark Company, 1951), 10.
6. U. S. Department of Commerce, *Materials Survey*, II–7.
7. R. L. Duffus, "Aluminum for War and for Peace," *New York Times Magazine*, 25 April 1943.
8. Aluminum Limited, *Aluminum Panorama* (Montreal: Aluminum Limited, 1953), 13. In 1934, before the advent of the World War II fighter plane, Lewis Mumford foreshadowed the important role of aluminum in any war effort, hailing the material as epitomizing the two important qualities of the neotechnic era: lightness and compactness. Mumford wrote, "And just as the technique of water-power and electricity had an effect in reorganizing even the coal-consumption and steam-production of power plants, so the lightness of aluminum is a challenge to more careful and more accurate design in such machines and utilities [that] still use iron and steel." *Technics and Civilization* (San Diego: Harcourt Brace & Company, 1934), 231. Mumford also prefigured the transformation of aluminum to the domestic realm: "Everything from type-writer frames to airplanes, from cooking vessels to furniture, can now be made of aluminum and its stronger alloys." (230).
9. U.S. Department of Commerce, *Materials Survey*, VIII–11.
10. Duffus, "Aluminum for War and for Peace," 8.
11. U.S. War Production Board, *American Industry in War and Transition*, vol. 2, 1945, 7.
12. Duffus, "Aluminum for War and for Peace," 8. Apart from aircraft needs, other military uses of aluminum included tank and automotive parts, packaging, landing mats, barracks, bridges, explosives, ammunition containers, ship fittings, and minesweeping cable fuses. Aluminum foil would continue to be an important element in the manufacturing of radar, anti-radar, and packaging. See also U.S. Department of Commerce, *Materials Survey*, VII–1.
13. John McDonald, "The War of Wits," *Fortune*, March 1951, 150.
14. Vannevar Bush, *Science: The Endless Frontier* (Washington, D.C.: U.S. Government Printing Office, 1945). One of the first OR programs implemented by the U.S. military was practiced at the young Anti-Submarine Warfare Operations Research Group, where a study of Allied submarine tactics concluded that the most effective use of depth-charges occurred when the depth of the charges would be alternated, increasing the probability of bombing a submerged enemy

submarine. Smith, *The RAND Corporation*, 6–7.

Another interesting early example of OR at work took place in Great Britain in 1942. The military studied airborne attacks on U-boats. By simple observation in the field, the Whitley airplane's black paint was sighted 12 seconds earlier by a submarine lookout than a plane painted white. As a result of this research, all Whitleys were painted white, thereby increasing the success rate of sinkings by one-third. See Robert Buderi, *The Invention that Changed the World: How a Small Group of Radar Pioneers Won the Second World War and Launched a Technological Revolution* (New York: Simon and Schuster, 1996), 149.

15. Buderi, *The Invention*, 149.
16. McDonald, "The War of Wits," 148.
17. Smith, *The RAND Corporation*, 35.
18. Ibid., 37.
19. Ibid., 36–37.
20. McDonald, "The War of Wits," 148.
21. Saul Friedman, "The RAND Corporation and Our Policy Makers," *Atlantic Monthly* 212 (Sept. 1963): 62.
22. The name *RAND* is an acronym for the phrase *Research ANd Development*.
23. It is important to note that the founding of what was to become the military-industrial complex was funded by the government without the need for congressional approval and without taking bids for the contract. The contract with Douglas Aircraft for Project RAND was signed by General Arnold. Friedman, "The RAND Corporation and Our Policy Makers," 62.
24. Rand Corporation, *The RAND Corporation: The First Fifteen Years* (Santa Monica, Calif.: RAND, 1963), 2.
25. Ibid., 9–10.
26. McDonald, "The War of Wits," 101.
27. Friedman, "The RAND Corporation and Our Policy Makers," 62.
28. McDonald, "The War of Wits," 101.
29. Ibid.
30. RAND Report SM–11827, Preliminary Design of an Experimental World-Circling Space Ship, 5–2–46. Also cited in Rand Corporation, *The RAND Corporation*, 9.
31. Friedman, "The RAND Corporation and Our Policy Makers," 66. Emphasis added.
32. U.S. Department of Commerce, *Materials Survey*, II–13.
33. U.S. War Production Board, *American Industry*, 2.
34. Nathaniel H. Engle, Homer E. Gregory, and Robert Mossé, *Aluminum: An Industrial Marketing Appraisal* (Chicago: Richard D. Irwin, Inc., 1945), 189, cited in Dennis Doordan, "Promoting Aluminum: Designers and the American Aluminum Industry," in *Design History: An Anthology*, ed. Dennis Doordan (Cambridge, Mass.: MIT Press, 1995), 160.
35. Merton J. Peck, *Competition in the Aluminum Industry, 1945–1958* (Cambridge, Mass.: Harvard University Press, 1961), 148.
36. George David Smith, *From Monopoly to Competition: The Transformations of Alcoa, 1888–1986* (Cambridge: Cambridge University Press, 1988), 311.
37. Ibid., 288.
38. Alcoa's status as a monopoly was officially ended during World War II when the government saw the monopoly as endangering the war effort (and therefore, human lives). The government was required to intervene and subsidize the creation of new plants, which were then handed over to newly established aluminum companies such as Reynolds and Kaiser. See Charlotte Muller, "The Aluminum Monopoly and the War," *Political Science Quarterly* 60, no. 1 (March 1945): 14–43.
39. Smith, *From Monopoly to Competition*, 289.
40. Ibid., 289.
41. Ibid. Doordan claimed that this market realization culminated, for the aluminum industry, in the design era. Doordan, "Promoting Aluminum," 160.
42. W. Phillips Davison, *U.S. Wartime Propaganda: The Role of the Propaganda Planner*, U.S. Air Force, Project RAND, (August 29, 1950), 3.

43. Ibid., 37.
44. Margaret B. W. Graham and Bettye H. Pruitt, *R & D for Industry: A Century of Technical Innovation at Alcoa* (Cambridge, Mass.: Cambridge University Press, 1990), 214.
45. Aluminum Company of America, "Review of Aluminum Markets," 31 December 1947, 2, box no. 108, folder 9, Marketing and Sales Series, Alcoa Archives, Library and Archives Division, Historical Society of Western Pennsylvania, Pittsburgh. Alcoa also designed and built a proto-typical prefabricated house constructed from as many aluminum parts as possible; the model, however, was never put into mass production. Aluminum became a likely material for such uses as roofing and siding, sash and trim, awnings, curtain walls and spandrels, builders' hardware, duct work and related applications, insect screening, wire, and cable electrical conductors. See also Robert Friedel, "Scarcity and Promise: Materials and American Domestic Culture during World War II," in *World War II and the American Dream: How Wartime Building Changed a Nation*, ed. Donald Albrecht (Cambridge, Mass.: MIT Press and National Building Museum, 1995), 80.
46. William B. Harris, "The Splendid Retreat of Alcoa," *Fortune* 52 (October 1955): 114. The first office building to be outfitted with an aluminum curtain wall was the Equitable Savings and Loan Building in Portland, Oregon of 1948, by Pietro Belluschi. See William Dudley Hunt, Jr., ed., *Office Buildings: An Architectural Record Book* (New York: F. W. Dodge Corporation, 1961), 10.
47. Martin, *The Organizational Complex*, 103. See also Reinhold Martin, "Atrocities. Or, Curtain Wall as Mass Medium," *Perspecta 32: Resurfacing Modernism*, 2001, 69.
48. Ketchum, MacLeod, and Grove, FORECAST Program Report/Scrapbook, 1959, on loan by Torrance Hunt, Acct. # L98.011, Alcoa Archives, Library and Archives Division, Historical Society of Western Pennsylvania, Pittsburgh.
49. Frank L. Magee, "Alcoa and Design," *Design Forecast* I (1959), 4, Publication Series, Alcoa Archive, Library and Archives Division, Historical Society of Western Pennsylvania, Pittsburgh. Emphasis added.
50. Ketchum et al., FORECAST Program Report/Scrapbook, 1959.
51. Coincidentally, the theme—along with the name of the program, FORECAST—corresponded with the Advertising Council's "The Future of America" national campaign, endorsed by and successfully conducted for the Eisenhower administration.
52. Herman Kahn and Anthony J. Wiener, *The Year 2000: A Framework for Speculation on the Next Thirty-Three Years* (New York: Macmillan Co., 1967), 6. See also Ian H. Wilson, "Scenarios," in *Handbook of Futures Research*, ed. Jib Fowles (Westport, Conn.: Greenwood Press, 1978), 228.
53. Ibid.
54. Herman Kahn, *Thinking about the Unthinkable* (New York: Horizon Press, 1962), 143. Kahn wrote this treatise on military strategy as a Research Associate of Princeton University while on leave from the RAND Corporation.
55. Stanford Research Institute, Center for the Study of Social Policy, *Handbook of Forecasting Techniques* (Fort Belvoir, Va.: U.S. Army Corp of Engineers, 1975), 193.
56. Alcoa, aluminum ball gown advertisement, featured in *The New Yorker*, October 1956.
57. Alcoa, summer house advertisement, featured in *The New Yorker*, May 1957.
58. For more on the national park system and the interstate highways that connected them, see Jeannie Kim, "Mission 66: Mission Control," on pages 168–89 in this volume.
59. Oscar Shefler, "What Kind of Nothing? The Unfettered Mind: Charles Eames and His FORECAST Solar Toy" in

Design Forecast I (1959): 38. Students from the California Institute of Technology visited Eames' studio to learn how he achieved his method of converting solar energy.

60. Shefler, "What Kind of Nothing?" 39.
61. Alcoa, view box advertisement, featured in *The New Yorker*, October 1959.
62. Samuel L. Fahnstock, "Cloud Nine," *Design Forecast* II (1960): 70–71, Publication Series, Alcoa Archives, Library and Archives Division, Historical Society of Western Pennsylvania, Pittsburgh.
63. Magee, "Alcoa and Design," 4.
64. The technical articles that appeared in the two issues of *Design Forecast* include "Mesh-Mash" (demonstrating new and experimental patterns of aluminum mesh); "Impacts—Inspiration for the Designer"; "Joining" (a photo exposé on different ways of joining aluminum); "Alloys for the Designer"; "A Forecast Metal"; "Castings—Inspiration for the Designer"; "Aluminum as Container"; and "Finishes for the Designer."
65. Joseph Petrocik, "Design Trends: Aluminum in Furniture," *Design Forecast* I (1959): 17. Some of the articles on specific designers included Shefler, "What Kind of a Nothing?"; Idem, "The Home that Is Not Home: Design for the 'Second Home,' Expressed in Robert Fitzpatrick's FORECAST Beach House Project"; Ilonka Karasz, "Inspiration: A Conversation and Graphic Essay"; and Saul Bass, "Some Thoughts on Motion Picture Film."
66. C. K. Rieger, "Appliances for Tomorrow—The Manufacturer's Challenge," *Design Forecast* I (1959): 50.
67. Walter Dorwin Teague, "When Should the Industrial Designer Enter the Picture?" *Design Forecast* II (1960): 45.
68. Ibid., 46. Teague elaborated on the concept of developing a corporate image by providing a few examples: "Olivetti in Italy has accomplished this identification superbly without any stereotype or repetitiousness whatever, with only an inexhaustible freshness that has become instantly recognizable. We are seeing this happen in a distinguished way with I.B.M. It is happening to Alcoa and to U.S. Steel."
69. The structure of the symposium mimicked the type of powerful yet enlightened "corporate retreat" atmosphere that Walter Paepke of the Container Corporation of America created in 1950 at the Aspen Institute and, in 1951, at the International Design Conference, the latter established in conjunction with Egbert Jacobson. The Aspen Conference, along with the American Assembly created by General Dwight Eisenhower at Columbia University, encouraged new types of business discourse. In an atmosphere of "relaxed authority," important decisions were contemplated by management types and members of the power elite, sometimes in a rustic outdoor setting, mimicking the brainstorming Delphi sessions conducted at government think tanks. Laurence S. Sewell, "Let's Start Where We Start: A Symposium," *Design Forecast* II (1960): 4.
70. "The word Delphi refers to the hallowed site of the most revered oracle in ancient Greece. Apollo... made himself master of Delphi. He was famous throughout Greece not only for his beauty, but also for his ability to foresee the future. The home Apollo chose for himself served not only as an oracular center but also as a kind of art museum." Harold A. Linstone, "The Delphi Technique," in *Handbook of Futures Research*, ed. Jib Fowles (Westport, Conn.: Greenwood Press, 1978), 273.
71. Ibid., 273. This highly-sensitive, early forecasting work was not published until 1964, with the RAND paper "Report on a Long-Range Forecasting Study." The time frame associated with the term "long-range" spanned from ten to fifty years. During the 1950s and 1960s, the U.S. Air Force conducted two comprehensive forecasting studies:

the Woods Hole Study and Project Forecast. The Woods Hole Study researched issues such as the design of aircraft and missiles, propulsion, space problems, medicine, materials, and behavioral sciences. Project Forecast involved both Air Force and industry and studied topics such as threat definition, technology development, and cost analysis forecasted for the next fifteen to twenty years. See Erich Jantsch, *Technological Forecasting in Perspective* (Paris: Organization for Economic Co-operation and Development Publications, 1967), 308.
72. Sewell, "Let's Start Where We Start," 4.
73. Douglas Kelley, quoted in Ibid., 6.
74. Bill Snaith, quoted in Ibid, 14–15.
75. Ibid., 16.
76. In her 1994 book *Privacy and Publicity: Modern Architecture as Mass Media* (Cambridge, Mass.: MIT Press), Beatriz Colomina has made the same claim regarding modern media and its evolution from the post–World War I years, where media technologies, along with transportation technologies such as automobiles and airplanes, had "emerged from the prewar revolution": "The media were developed as part of the technology and instrumentation of war. . . . After the war, this technology was gradually domesticated. Just as regular airline services were being established throughout Europe at the beginning of the twenties, radios and telecommunications had become household items" (156).
77. Brook Stevens, head of the design firm Brook Stevens Associates, quoted in Sewell, "Let's Start Where We Start," 17.
78. Smith, *From Monopoly to Competition*, 312.
79. John Harper, *Design Forecast* (brochure), Aluminum Company of America, December 1965, MSS box no. 138, folder 5, Marketing and Sales Series, Alcoa Archives, Library and Archives Division, Historical Society of Western Pennsylvania, Pittsburgh.
80. Ibid., 4.
81. Ibid.
82. Ibid., 6.
83. Ibid., 10.
84. Ibid., 1.
85. Paul Edwards, *The Closed World: Computers and Politics of Discourse in Cold War America* (Cambridge, Mass.: MIT Press, 1996). Edwards' book contains an interesting historical account of the development of the electronic field during the Vietnam War with the implementation of systems analysis and operations research.

PLASTICS

1. Betty Pepis, "Plastics Limned in 'Dream Houses,'" *The New York Times*, 29 October 1952.
2. Ibid.
3. Robert Mueller, quoted in ibid.
4. Frank Curtis, "Monsanto in World War II—Summary of Division and Plant Reports Written in 1945," 15 April 1951, 1, Monsanto Company History World War II, box 4, series 10, Monsanto Historic Archive Collection, Washington University Library, St. Louis, Mo.
5. Ibid.
6. Ibid.
7. Ibid., 14. See also "Development Projects which were Either Completed by or Received the Attention of Monsanto during the War," Monsanto Company History World War II, box 4, series 10, Monsanto Historic Archive Collection, Washington University Library, St. Louis, Mo.
8. Curtis, "Monsanto in World War II," 15 April 1951, 15, Monsanto Historic Archive Collection.
9. Ibid.
10. Ibid.
11. As an exception to the rule, Dunlop observed that "there was a group in the A.S.T.M. [American Society of Testing Materials] that had seen that some day definite specifications and test methods to supplement these specifica-

tions would be necessary." He also noted, however, that "this group worked diligently without great encouragement." R. D. Dunlop, quoted in ibid.

12. Edgar Queeny et al., "Monsanto Chemical Company's Part in the War Effort: Report to George D. Hansen," Price Adjustment Section, Chemical Warfare Service, Washington, D.C., 5 February 1943, 5, Monsanto Company History World War II, box 4, series 10, Monsanto Historic Archive Collection, Washington University Library, St. Louis, Mo.
13. Monsanto manufactured Polyvinyl Buvtar for raincoats, gas-protective clothing, hospital sheeting, and military packaging; melamine resins of high resistance for aircraft ignition parts that were exceptional for replacing aluminum in dishware; Resinox with high impact strength for gunstocks, motor shell fuses and mines, tank periscopes, and bomb nosepieces; and Styrene with non-flammable, heat-resistant properties for the synthetic rubber program. Queeny et al., "Monsanto Chemical Company's Part," 3–6.
14. Hubert Kay, "Monsanto Products Used in World War II," Monsanto Company History World War II Products War Related, box 4, series 10, Monsanto Historic Archive Collection, Washington University Library, St. Louis, Mo.; "Doron: A Now-It-Can-Be-Told about Plastic Armor for American Troops," *Monsanto Magazine* 22–25 (October 1945): 34.
15. "The Nose that Sees," *Monsanto Magazine* 34, no. 4 (July–August 1954): 15–17.
16. Marvin E. Goody et al., *Building with Structural Sandwich Panels*, ed. Bernard P. Spring (Cambridge, Mass.: Massachusetts Institute of Technology, Department of Architecture, 1958), 7. Rotch Library, Massachusetts Institute of Technology, Cambridge, Mass.
17. The latter were manufactured by Guillespie Furniture Company of Los Angeles. Fred Galen, "Extending Our Sting," *Monsanto Magazine* 24, no. 3 (June 1945): 18–19.
18. Ibid.
19. "Baked to Order in 8 Minutes," *Monsanto Magazine* 23, no. 6 (November 1944): 13.
20. Ibid.
21. Teflon plastic was discovered by DuPont in their search for a material that might protect against fluorine gas, an extremely corrosive substance used for gaseous diffusion for atomic weaponry. Teflon was used to protect valves and gaskets needed for manufacturing the atomic bomb. See Stephen Fenichell, *Plastic: The Making of a Synthetic Century* (New York: Harper Collins, 1996), 221.
22. "Test-Tube Marvels of Wartime Promise a New Era in Plastics," *Newsweek*, 17 May 1943, 42.
23. "Plastics Tomorrow," *Scientific American* 170 (March 1944): 105.
24. "Test-Tube Marvels," 42.
25. "Indestructible Room: New Plastics Protect Walls, Furniture and Rugs from Ravages of Kids and Dogs," *Life*, 14 January 1946, 91; Christine Holbrook and Walter Adams, "Dogs, Kids, Husbands: How to Furnish a House so They Can't Hurt It," *Better Homes and Gardens* 27 (March 1949): 37.
26. "Plastics: A Way to a Better More Carefree Life," *House Beautiful* 89, no. 2 (October 1947): 120, 122–23.
27. Ibid., 138.
28. Ibid., 125.
29. Jeffrey L. Meikle, *American Plastic: A Cultural History* (New Brunswick: Rutgers University Press, 1995), 196. Meikles's text is a significant resource on the subject of plastics.
30. Ibid., 197.
31. Thomas Hines, *Populuxe* (New York: Knopf, 1986), 97.
32. Tensile strength construction techniques found significant application in furniture design before the war. The submissions of Charles and Ray Eames and Eero Saarinen to the 1940 Organic Design in Home Furnishings Competition at the Museum of Modern Art in New York were

eventually developed by the Eameses for use in leg splints and molded aircraft sections during World War II.

33. The Eameses originally experimented with fiberglass plastics in their Case Study House 8, sponsored by *Arts & Architecture* magazine between 1945 and 1949.
34. "Furniture for Moderns," *Monsanto Magazine* 34, no. 1 (January–February 1954): 15–16.
35. Dan Forrestal, *The Story of Monsanto: Faith, Hope, and $5,000. The Trials and Triumphs of the First Seventy-Five Years* (New York: Simon and Schuster, 1977), 143.
36. Ralph Hansen, "Plastics in the Design of Building Products and Their Markets," quoted in R. K. Mueller, "Confidential—First Disclosure: The Plastics Division Presents the House of Tomorrow," 3 October 1955, 6–7, box 3, series 9, Monsanto Historic Archive Collection, Washington University Library, St. Louis, Mo.
37. Hansen, quoted in ibid.
38. Ibid., 3.
39. Ibid. Monsanto's confidential report maintained that annual average new households per year would drop from 1,525,000 between 1947 and 1950 to only 818,000 between 1950 and 1954. They cited the U.S. Census Bureau's projection of a further decline to 630,000 annual average households per year between 1955 and 1960. After 1960, annual average households were projected to increase to 1,500,000.
40. Hansen, quoted in ibid, 6–7.
41. Ibid., 7.
42. Richard Hamilton et al., *Architectural Evolution and Engineering Analysis of a Plastics House of the Future* (Cambridge, Mass.: Massachusetts Institute of Technology, Department of Architecture, 1957), 1. Rotch Library, Massachusetts Institute of Technology, Cambridge, Mass.
43. Miekle, *American Plastic*, 205.
44. *Plastics in Housing* (Cambridge, Mass.: Massachusetts Institute of Technology, Department of Architecture, 1955), 1–56, limited Access Collection, Rotch Library, Massachusetts Institute of Technology, Cambridge, Mass.
45. Betty Pepis, "People in Plastic Houses," *New York Times Magazine*, 23 November 1954, 50–51.
46. Goddy et al., *Building with Structural Sandwich Panels* (Cambridge, Mass.: Massachusetts Institute of Technology, Department of Architecture, 1958), 8. Rotch Library, Massachusetts Institute of Technology, Cambridge, Mass.
47. Ibid., iii.
48. "P/A News Survey: Monsanto Reveals Present and Future of Plastics in Architecture," *Progressive Architecture*, June 1957, 89.
49. Goddy et al., *Building with Structural Sandwich Panels*, 3.
50. "Big Impact, Bigger Potential," *Newsweek*, 4 June 1956, 80.
51. Mueller, "Confidential—First Disclosure," 10.
52. Douglas Haskell, "In Architecture, Will Atomic Processes Create a New 'Plastic' Order?" in "Building in the Atomic Age," *Architectural Forum*, September 1954, 100.
53. Haskell, "In Architecture," 100. Emphasis in the original.
54. Kiesler's Endless Theater project, conceived between 1924 and 1926 and displayed at the International Theater Festival in New York in 1926, originally presented the concept of dual-shell glass and steel egg-shaped structural skin.

    Kiesler first articulated continuous tension shell construction for domestic architecture in his Space House project, manufactured for the Modernage Furniture Company of New York in 1933. Kiesler's structural concept was informed by innovative bridge design. Eugéne Freyssinet, the French engineer, and Robert Maillart, the Swiss bridge builder, explored tension shell construction using steel reinforced concrete in the 1920s and 1930s. The structural concepts presented by Douglas Haskall and explored by MIT for MHOF were originally conceived for glass, concrete, and steel composite materials and adapted to plastics.

55. "Does Atomic Radiation Promise a Building Revolution?" in "Building in the Atomic Age," *Architectural Forum*, September 1954, 94.
56. Ibid., 96.
57. Dr. Charles A. Thomas, the director of central research and vice president of Monsanto, had completed vital research and solved production problems as director of the Clinton Tennessee Laboratories working on radioactive isotopes. Monsanto took control over the Clinton Laboratories from the University of Chicago in 1945 and also assumed construction and supervision of the Atomic Energy Commission lab in Miamisburg and Marion, Ohio. Monsanto was an active force in the development of atomic research and production, receiving numerous wartime and postwar contracts. See Dr. M. D. Whitaker, "Manhattan Project, Oak Ridge, Tennessee—Name of Dogpatch," in "Nucleonics," *Monsanto Magazine* 24, no. 6 (1945): 16. See also Curtis, "Monsanto in World War II," 20.
58. Haskell, "In Architecture," 100. Emphasis in original.
59. *Monocoque* is an aeronautical engineering term for a structure, such as an airplane fuselage, that has an outer covering in the form of a rigid skin or shell designed to bear all or most of the structural stresses.
60. Program Brochure, Massachusetts Institute of Technology 1955 Summer Conference on Plastics in Housing, quoted by Ralph Hansen in Mueller, "Confidential—First Disclosure," 2.
61. R. C. Evan, quoted in "News for Release—Immediately," Press Book, Monsanto Chemical Company, Springfield, Massachusetts, June 1956, Monsanto Company History, box 3, series 9, Monsanto Historic Archive Collection, Washington University Library, St. Louis, Mo.
62. "General Background Information, Monsanto Chemical Company 'House of the Future,'" 1957, 4, Exhibits and Visual Arts (House of the Future), box 3, series 9, Monsanto Historic Archive Collection, Washington University Library, St. Louis, Mo.
63. Hamilton et al., "Architectural Evolution," i.
64. Mueller, "Confidential—First Disclosure,"1.
65. Hamilton et al., "Architectural Evolution," 2.
66. Ibid.
67. Ibid.
68. Ibid.
69. Ibid.
70. Ibid.
71. A. G. H. Dietz et al., "Engineering the Plastics 'House of the Future,' Part I," *Modern Plastics*, June 1957, 146.
72. Hamilton et al., "Architectural Evolution," 3.
73. Ibid.
74. Ibid., 2.
75. Ibid., 6.
76. Dietz et al., "Engineering the Plastics 'House of the Future,'" 150.
77. Ibid.
78. Hamilton et al., "Architectural Evolution," 6.
79. Dietz et al., "Engineering the Plastics 'House of the Future,'" 148.
80. Hamilton et al., "Architectural Evolution," 6.
81. Ibid., 5.
82. Ibid., 21.
83. Dietz et al., "Engineering the Plastics 'House of the Future,'" 150. As plastics of the time had a thermal expansion two to eight times as large as other construction materials, the differential between the heat of the sun and the air-conditioned interior created uneven distribution of buckling forces, which caused twisting. Wind forces also created twisting and uplift, as did earthquake forces.
84. Dietz et al., "Engineering the Plastics 'House of the Future,'" 150.
85. "Plastics—Shaping Tomorrow's Houses?" *Architectural Record*, August 1956, 210.

86. A. G. H. Dietz et al., "Engineering the Plastics 'House of the Future,' Part II," *Modern Plastics*, July 1957, 127.
87. "Monsanto's House of Tomorrow," *Monsanto Magazine* 36, no. 4 (August–September 1956): 17.
88. "Like most explorers on the frontier of the unknown, Monsanto is 'traveling light.' That is, at this time, it does not plan to make or sell duplicates of the House of Tomorrow." Ibid.
89. R. K. Mueller, Monsanto Chemical Company, St. Louis, to Mr. Persechini, 6 October 1955, 2, Monsanto Company History, box 3, series 9, Monsanto Historic Archives Collection, Washington University Library, St. Louis, Mo.
90. Monsanto Chemical Company, Plastics Division, "News for Release—Immediately," Press Book, Monsanto Chemical Company, Springfield, Massachusetts, June 1956, 2, Monsanto Company History, box 3, series 9, Monsanto Historic Archive Collection, Washington University Library, St. Louis, Mo.
91. Monsanto Chemical Company, Plastics Division, "Project Effectiveness to Date," *Directory: The Monsanto "House of The Future,"* Summer 1957, 21, Executive Report, box 1, series 9, Monsanto Historic Archive Collection, University of Washington, St. Louis, Mo.
92. "Disneyland: Fabulous Wonderland Opens in Anaheim; Lath and Plaster Used Extensively in Producing Structures for Show Place," *The California Plasterer*, July 1955, 15, Walt Disney Archives, Burbank, Calif.
93. Karal Ann Marling, "Imagineering the Disney Theme Parks," in *Designing Disney's Theme Parks*, ed. Karal Ann Marling (Montreal: Canadian Centre for Architecture, 1997), 143.
94. Jack E. Janzen, "The Monsanto Home of the Future: Putting the 'Tomorrow' in Tomorrowland," in *The "E" Ticket* (Winter 1991–92): 12, 15, Walt Disney Archives, Burbank, Calif.
95. See C. G. Cullen, "Fabricating the Structural Components of the All-Plastic 'House of the Future,'" *Plastics Technology* 10 (October 1958): 921–27. See also Meikle, *American Plastic*, 209–10, for more on fabrication techniques.
96. Gladwin Hill, "Four Wings Flow from a Central Axis in All-Plastic 'House of Tomorrow,'" *The New York Times*, 12 June 1957.
97. Hill, "Four Wings Flow." See also Thomas Bush, "Push Button, Pine-scented Plastic House with 'Floating' Rooms Shown at Disneyland," *The Wall Street Journal*, 13 June 1957, 5.
98. Monsanto Chemical Company, "Press Book," in *Directory, The Monsanto "House of the Future,"* 17 July 1955, Executive Report, box 1, series 9, Monsanto Historic Archive Collection, University of Washington, St. Louis, Mo.
99. Monsanto Plastics Division, "A 1957 Summary and Analysis, and 1958 Forecast of MHOF Publicity," Special to D. J. Forrestal, September 1957, 1, Monsanto Company History, box 3, series 9, Monsanto Historic Archive Collection, Washington University Library, St. Louis, Mo.
100. "A Disneyland Study for Monsanto Chemical Company Hall of Chemistry and House of the Future," Customer Relations Division, Disneyland, Inc. for Monsanto Chemical Company, October 1958, Monsanto House of the Future folder, Walt Disney Archives, Burbank, Calif.
101. Monsanto Plastics Division, "News for Release—Immediately," Press Book, Monsanto Chemical Company, New York, New York, 1958, 1–2, History, box 3, series 9, Monsanto Historic Archive Collection, Washington University Library, St. Louis. Mo.
102. "Right to the End, This Plastic Structure Proved that It Was a House with a Bounce," *Monsanto Magazine*, March 1968, 25.
103. "Right to the End," 25.
104. Ibid., 23.

105. Ibid., 25.
106. Ibid., 25.
107. Albert G. H. Dietz, "Is a Plastics Breakthrough in Building Due in the Sixties?" *Architectural and Engineering News* 2 (July 1960): 4.

PLAYROOM

1. Benjamin Spock, M.D., "What We Know about the Development of Healthy Personalities in Children," *Mid-Century White House Conference on Children and Youth* 1 (Washington, D.C.: typed manuscript, 1950), 5.
2. Hannah Arendt, *The Human Condition* (Chicago: University of Chicago Press, 1958), 127.
3. Ibid., 128.
4. Anne Kelley, "Suburbia—Is It a Child's Utopia?" *The New York Times Magazine*, 2 February 1958, reprinted as "For the Sake of the Children" in *Suburbia in Transition*, ed. Louis H. Masotti and Jeffrey K. Hadden (New York: New View Points, 1974), 21–27.
5. Scott Donaldson, *The Suburban Myth* (New York: Columbia University Press, 1969), 135. Although Anne Kelly and others attributed the focus on children in postwar suburban culture to the fact that so many children were born following World War II, the demographic surplus of children was only one critical element in a series that shaped the social environment of postwar America.
6. Dieter Lenzen, "Disappearing Adulthood: Childhood as Redemption," in *Looking Back on the End of the World*, ed. Dietmar Kamper and Christoph Wulf, trans. David Antal (New York: Semiotext(e), 1989), 65.
7. "Planning for Complete Flexibility," *Architectural Forum*, April 1950, 128.
8. Ibid.
9. David Riesman, "The Suburban Dislocation," *Annals*, November 1957, 130–42.
10. Hayden Phillips in "Planning for Complete Flexibility," 127–31.
11. "Builder and Architect," *Architectural Forum*, April 1950, 117.
12. Ibid.
13. Ibid., 118.
14. "Quality Houses through Contemporary Design," *Architectural Forum*, April 1950, 122.
15. Ibid., 123.
16. Ibid. "To obtain the maximum use of the limited space in today's small house the contemporary architect makes the same square footage serve numerous purposes. Thus, between meals the dinning area may be added to the living area or double as an entry hall or the children's play space."
17. This is accomplished "mainly through the provision of large glass areas. Big windows bring the daylight, the garden and the more distant view into the house and relieve the rooms of their cell-like boxiness." Ibid.
18. "Planning for Complete Flexibility," 128.
19. The novelty of Phillips' work is in its purported "scientific/statistical" basis and his method of representation. The reliability of his technique and accuracy of his data, however, must be questioned.
20. "Planning for Complete Flexibility," 128.
21. Tailor-Made Houses," *Architectural Forum*, April 1950, 172.
22. Ibid., 118.
23. *Architectural Record*, May 1950, cover. "Here we are concerned not only with design, materials, equipment and construction; concurrently we are reporting on the small house as a field of architectural practice. The demand is tremendous; in 1948, 766,500 one-family houses, averaging $7,850 in cost, were started." Ibid., 125. This articulation of the issue as a "field of architectural practice" reflects the underlying editorial position that suburban tract-housing design—since the war, mainly the domain of builders and developers—would benefit greatly from the increased involvement of design professionals.
24. TAC partners initially included Jean Bodman Fletcher, Norman Fletcher,

Walter Gropius, John C. Harkness, Sarah Harkness, Robert S. McMillan, Louis A. McMillan, and Benjamin Thompson.

25. The winners in the third year of competition sponsored by the Revere Quality House Division of the Southwest Research Institute were selected from across the U.S. To earn the Quality House seal of approval, an architect-designed house costing less than $20,000 and located in a subdivision needed to be built to certain minimum design and materials standards. The builder was required to guarantee construction for a year. "Building Types Study No. 161," *Architectural Record*, May 1950, 136.
26. "Six Moon Hill," *Architectural Record*, June 1950, 113.
27. Ibid.
28. Ibid.

TOY

1. "Building Toy," *Life*, 16 July 1951, 58.
2. Ibid., 60.
3. These films included *Traveling Boy*, 11:45 min., color, 1950; *Parade*, 5:33 min., color, 1952; *Tops*, or *Stars of Jazz*, 3:01 min., black and white, 1957; and *Toccata for Toy Trains*, 13:28 min., color, 1957. All films were produced by the Office of Charles and Ray Eames, Los Angeles, Calif.
4. Charles Eames, narration of opening segment of the film *Toccata for Toy Trains*, quoted in John Neuhart, Marilyn Neuhart, and Ray Eames, *Eames' Design* (New York: Harry N. Abrams), 215.
5. *The Toy* was manufactured by Tigrett Enterprises in Jackson, Tenn. and was included in the Sears Roebuck catalog for a few seasons. In 1951 it cost $3.50. Tigrett enterprises went out of business in 1961, and *The Toy* was not manufactured again afterward.
6. Publicity leaflet for *The Toy*, box 229, 1951, Administrative/Publicity File, The Works of Charles and Ray Eames, Library of Congress.

   An earlier working prototype of *The Toy* was comprised of only triangles. The Eameses added the rectangular elements to speed up the construction process and to enable players to build larger, taller forms.

   Also manufactured by Tigrett Enterprises was *The Little Toy*, which appeared a year after *The Toy* was launched. A smaller version of *The Toy*, *The Little Toy* came with wire frames in the shape of triangles and squares that could be assembled to create row houses, bridges, or rocket-launching platforms. Colorful cardboard panels took the place of *The Toy*'s paper panels and could be connected to the frames. Produced until 1961, *The Little Toy* was advertised as creating small environments to play around, while *The Toy* was advertised as providing larger environments to play within.
7. *Flexikite*, for example, was manufactured between 1950 and 1951 and came in a cardboard tube. It was invented and patented by Francis and Gertrude Rogallo in November 1948. It had no rigid structure and led to the subsequent design of a parachute for NASA (Para Wing) by Francis Rogallo, an aerona utical engineer.
8. Letters of solicitation, box 104, The Works of Charles and Ray Eames, Manuscripts Division, Library of Congress.
9. Of the two hundred kites exhibited at the Hallmark Gallery in New York City during the summer of 1968, seventy-one were loaned by Charles and Ray Eames. A year later, the Eameses sent ninety-seven of their kites to the Field Museum of Natural History in Chicago for the exhibition *The Wind in My Hands*.
10. *Polavision* was produced for the Polaroid Corporation by the Office of Charles and Ray Eames in Los Angeles, Calif., in 1978. In the film, short vignettes, each two-and-a-half minutes long, demonstrated aspects of everyday life.
11. The Case Study House Program, sponsored by John Entenza and *Art and Architecture* magazine, enlisted young architects to design and build low-cost

innovative architecture with new materials. Case Study House #8 was designed by Charles and Ray Eames for their own use, while Case Study House #9, its neighbor, was designed and built at the same time by Charles Eames and Eero Saarinen for John Entenza. Both houses used steel-frame construction, but in the Entenza House (#9) no beams or columns were expressed. For more on the Case Study House Program, see Elizabeth A.T. Smith, ed., *Blueprints for Modern Living: History and Legacy of the Case Study House* (Los Angeles: Museum of Contemporary Art, and Cambridge, Mass: MIT Press, 1989). For more on the Eames House specifically, see Neuhart, Neuhart, and Eames, *Eames' Design*, 106–21.

12. "Life in a Chinese Kite," *Architectural Forum*, September 1950, 90.
13. Ibid.
14. Lawrence Hargrave (1850–1915) invented his kite in Australia in 1892 and put it to use in meteorological experiments. See Alexander Graham Bell, "The Tetrahedral Principle in Kite Structure," *National Geographic Magazine*, June 1903, 219–50.
15. The *Meccano Set* was the British predecessor of the American *Erector Set*, which was manufactured by the A. C. Gilbert Company, New Haven, Conn., between 1913 and 1962. Both toys were comprised of steel parts that would connect with screws. *Meccano*'s parts were mainly perforated steel rods, while *Erector*'s were small steel girders.
16. Though Charles and Ray Eames are both credited with the design of the Eames House, this article mentions Charles Eames alone as its architect. "Life in a Chinese Kite," 93–94, 96.
17. Ibid., 96. "One of Eames' many surprise discoveries as the house went up was that light steel is a distinct material, very different from its familiar, heavy parent."
18. Ibid. In summing up his work in the article, Charles Eames mentions his regret that he had not been influenced to a greater extent by the marine and aviation equipment with which he was familiar in the choice of materials for the house.
19. For more on the Eames' practice of combining and recombining elements in their designs, see Beatriz Colomina, "Reflections on the Eames House," in *The Work of Charles and Ray Eames: A Legacy of Invention* (New York: Harry N. Abrams, 1997), 126–49.
20. In light of the tremendous expenditure needed to mass-produce the Wichita House, at the time of the Cold War there was little incentive to retool the Beech Aircraft Company for this domestic purpose.
21. R. Buckminster Fuller, *Designing a New Industry* (Wichita, Kans.: Fuller Research Institute, 1946), 2.
22. The copy of *Designing a New Industry*, a pamphlet published upon the completion of the Wichita House, with the dedication to the Eameses, can be found in box 30, folder 9, Office Files, Fuller, R. Buckminster, The Works of Charles and Ray Eames, Manuscript Division, Library of Congress.
23. Ibid., 2.
24. Ibid., 3.
25. Alden Hatch, *Buckminster Fuller: At Home in the Universe* (New York: Crown Publishers, 1974), 153.
26. John McHale, *R. Buckminster Fuller* (New York: Braziller, 1962), 29.
27. R. Buckminster Fuller, quoted in Hatch, *Buckminster Fuller*, 153. Fuller explained, "all polyhedra may be subdivided into component tetrahedra, but no tetrahedron may be subdivided into component polyhedra of less than the tetrahedron's four faces... for we cannot find an enclosure of less than four sides." R. Buckminster Fuller and E. J. Applewhite, *Synergetics: Explorations in the Geometry of Thinking* (New York: MacMillan Publishing, 1975), 335.
28. "The tetrahedron is the basic structural system, and all structure in the universe

is made up of tetrahedronal parts." Fuller, quoted in Hatch, *Buckminster Fuller*, 153.

29. Fuller, according to McHale, *R. Buckminster Fuller*, 15.
30. Robert Marks, *The Dymaxion World of Buckminster Fuller* (Carbondale and Edwardsville: Southern Illinois University Press, 1960), 43.
31. Fuller assumed that the most economical structural energy web might be derived through the fusion of a tetrahedron and a sphere. The sphere encloses the largest amount of space with the smallest surface and is strongest against internal pressure, while the tetrahedron encloses the least amount of space with the largest surface and is strongest against external pressure. Fuller and Applewhite, *Synergetics*, 52–56, 319–30.
32. R. Buckminster Fuller, *Tetrascroll, Goldilocks, and the Three Bears* (1975; reprinted New York: St. Martin's Press, 1982).
33. Fuller, *Tetrascroll*, vii.
34. Wachsmann, together with Walter Gropius, founded the General Panel Corporation in September 1942 and developed a factory-made house with the aim of providing the postwar American market with industrialized suburban housing. The Packaged House engaged their attention between 1942 and 1951, and although it failed as a business endeavor, the architects' combined efforts created valuable prototypes for prefabricated housing. See Gilbert Herbert, *The Dream of the Factory-Made House: Walter Gropius and Konrad Wachsmann* (Cambridge, Mass.: MIT Press, 1984).
35. Konrad Wachsmann, *The Turning Point of Building* (New York: Reinhold Publishing Corporation, 1961), 170.
36. Ibid., 12.
37. Ibid., 9.
38. Ibid., 186.
39. Robert Le Ricolais, "Les Réseaux à trois dimensions: À propos du project de hangar d'aviation de Konrad Wachsmann," *Architecture d'Aujourd'hui* 25 (July–August 1954): 10–12.
40. Ibid.
41. Wachsmann reported that a team of students and consultants worked on his design, which was developed mainly through models rather than drafting. Wachsmann, *The Turning Point of Building*, 186.
42. Wachsmann, *The Turning Point of Building*, 22–31. Kenneth Frampton called Wachsmann and Fuller "the Technocrats of the Pax Americana" and wrote, "If there was a single techno-scientific genius admired by both men, it was surely Alexander Graham Bell. . . . Bell was seen by both men as the modern equivalent of the *uomo universale*." Frampton, "I tecnocrati della Pax Americana: Wachsmann & Fuller," *Casabella* 542–43 (January–February 1988): 40–45.
43. Alexander Graham Bell, "The Tetrahedral Principle in Kite Structure," *National Geographic Magazine* XIV, no. 6 (June 1903): 225.
44. Alexander Graham Bell, "Aerial Locomotion," *National Geographic Magazine* XVIII, no. 1 (Jan. 1907): 11.
45. Ibid. At the time, Bell was president of the National Geographic Society and published many of his findings in *National Geographic Magazine*. He headed a group of five men—the Aerial Experiment Association—that, funded by his wife, Mabel Hubbard Bell, carried out countless well-recorded attempts to fly at Beinn Bhreagh in Nova Scotia.
46. Bell, "Aerial Locomotion," 10.
47. Bell, "The Tetrahedral Principle," 225.
48. Ibid., 224. Furthermore, Bell observed that the flatter the wings, the more unstable the arrangement in the air, as, with an upset, one side would lift while the other would depress. The more the wings are raised, the more they approach a perpendicular position, the more stable the arrangement is in the air. The dividing line between the two conditions is at 45 degrees; thus,

the tetrahedron's angles, at 60 degrees, form a stable structure.

49. This was because the weight was increased to the power of three, while the surface, which was the kite's lifting vehicle, increased only to the square of the dimension.
50. Bell, "Aerial Locomotion," 12.
51. Ibid., 11.
52. Bell calculated that a structure twice the size, or 40 feet across, could sustain both a man and a motor while flying at a low velocity.
53. J. H. Parkin, *Bell and Baldwin* (Toronto: University of Toronto Press, 1964), 441. The Wright Brothers had flown their biplane successfully by 1903. Bell's probing of the idea of a plane made of tetrahedral cells was due in great part to his concern for the safety and stability of flight, which he felt the kite could increase. These reflections led him to the design of aerodromes (a term he preferred to aeroplanes) and the development of hydrodromes until his death in 1922. Parkin, *Bell and Baldwin*, 17–18.
54. Bell, "The Tetrahedral Principle," 231.
55. The tower stood for about a decade before it was dismantled. During that time it was said to require hardly any maintenance. The tower, which offered commanding views of the surrounding hills of Cape Breton and of Bras d'Or Lake, was formally opened with a ceremony on August 31, 1907, when guests, including women in long skirts, climbed over one hundred steps to reach the observation deck at its pinnacle. Parkin, *Bell and Baldwin*, 33.
56. Wachsmann, *The Turning Point of Building*, 31.
57. Johan Huizinga, *Homo Ludens: A Study of the Play Element in Culture*, 1st paperback ed., trans. R. R. C. Hull (Boston: Beacon Press, 1955), 8ff.
58. Ibid., 8.
59. Ibid.
60. Ibid., 10.
61. Ibid., 8.
62. For a discussion of triangulated space frames and their relation to networks of communication, see Mark Wigley, "Network Fever," *Grey Room* 4 (Summer 2001): 82–122.
63. "Atomic Bomb over Nevada," *Life*, 16 July 1951, 51.

MISSION 66

1. Alexander Wilson, "The View from the Road: Recreation and Tourism," in *The Culture of Nature: North American Landscape from Disney to the Exxon Valdez* (Cambridge: Blackwell Publishers, 1992), 44.
2. Ibid., 19–20.
3. Ibid., 20–23.
4. Although taking place outside the city, these leisure activities remained inextricably linked to the schedules and geographies of urban life. See Ibid., 19–44, for an entertaining account of the way that the road has shaped the American understanding of tourism and landscape.
5. Early demographic studies of the national parks reveal that the visitorship was heavily biased toward white, middle-class families. Despite occasional voices of dissension, little was done to ameliorate the seemingly unintentional segregation. See Joseph W. Meeker, "Red, White, and Black in National Parks," *The North American Review* 258, no. 1 (1973): 11–14, for additional sources.
6. Fraser Darling, "Man and Nature in the National Parks: Reflections on Policy," *National Parks Magazine* 43 (1969): 17.

   Through the efforts of Mission 66—and, according to its critics, as a result of its close relationship with the AAA—the NPS developed a system of drive-in campsites and other facilities dedicated to the automobile. These sites were seemingly devoid of remarkable natural features and were ecologically unsound. (Views taken from the air suggest that they were envisioned as an extension of the suburban planning model.)

   The tension between access—specifically, vehicular access—and ecology within the national park system would plague the agency throughout its history.

7. It is worth mentioning that the National Institute of Mental Health (NIMH) was founded in 1946 as a federal agency. The organization was originally established to combat narcotic addiction through treatment farms in Lexington, Kentucky and Fort Worth, Texas. Within a few years, the mandate of the NIMH was expanded to include nervous and mental diseases based upon the experiences of those in military service during World War II. By the mid-sixties, alongside the NPS in the Bureau of State Services, the NIMH began to pursue issues of water safety, air pollution, and environmental health before institutionalized environmental legislation was enacted during Nixon's administration. Although this is seemingly unrelated to our concerns here, the neighborliness of the NPS and the NIMH during the time of Mission 66 and the "curative" aspect described as part of the mission of the National Park system (in light of the increasing girth of the National Institute of Mental Health) are, perhaps, worthy of further consideration.
8. Conrad Wirth, in the preface to Roy E. Appleman, *A History of the National Park Service Mission 66 Program* (Washington, D.C.: Department of the Interior, 1958), 2.
9. Ibid., 29.
10. Ibid.
11. Audience Research, Inc., *A Survey of the Public Concerning the National Parks* (Princeton: Audience Research, Inc., 1955), 117.
12. U.S. Department of the Interior, *The Need for Mission 66* (Washington, D.C.: Government Printing Office, 1956), 11.
13. Department of Forest Economics, *Our Heritage* (Washington, D.C.: Government Printing Office, 1956), [3].
14. Ibid., inside cover.
15. U.S. Department of the Interior, *The Need for Mission 66*, 117.
16. Department of Forest Economics, *Our Heritage*, not paginated [1].
17. Ibid.
18. Department of Forest Economics, *Our Heritage*, [3].
19. Ibid., [24].
20. Threatened by the arrival of the NPS in 1916, the USFS frequently and vocally denounced the NPS as merely a tourist agency for America's natural resources, while promoting itself as a conservationist and ecologically-driven institution. See Ronald A. Foresta, *America's National Parks and Their Keepers* (Washington, D.C.: Resources for the Future, 1984) for a history of this early relationship as well as the influence of the Civilian Conservation Corps upon the policies of both institutions between the wars.
21. The caravan tour caused too much traffic and was phased out by 1932.
22. Barry Mackintosh, *Interpretation in the National Park Service* (Washington, D.C.: Government Printing Office, 1953), 22, 24.
23. Sal Prezioso, "A New Venture," *Journal of Leisure Research* 1, no. 1 (1968): vi. Roy E. Appleman, biographer of the Squad and future director of the National Park Service's interpretive efforts, responded with research. By the winter of 1969, he had helped organize an editorial advisory board charged with launching the *Journal of Leisure Research*. The board, which included members of the American Institute of Planners, the NPS, and various colleges with landscape architecture programs, embraced the contemporary penchant for modeling and predicting future events and conditions and embarked on a mission to apply algorithms and equations to predicting activity and desire within the park system. In its first year, the journal included articles on the maximum noise levels for optimum enjoyment in national parks, the possibility of the use of robots for park maintenance, the benefits of captive rearing of endangered species, and the philosophical development of leisure.

24. U.S. Department of the Interior, *Mission 66 for the National Park System* (Washington, D.C.: Government Printing Office, 1956), 29, 92. Ned J. Burns, *Field Manual for Museums* (Washington, D.C.: National Park Service, 1941), 24.
25. Sarah Allaback, *Mission 66 Visitor Centers: The History of a Building Type* (Washington, D.C.: U.S. Department of the Interior, 2000), 27.
26. Emerson Goble, "Architecture? for the National Parks," *Architectural Record*, January 1957, 121, as quoted in Allaback, *Mission 66 Visitor Centers*, 29.
27. Foresta, *America's National Parks*, 57, 59.
28. Allaback, *Mission 66 Visitor Centers*, 25.
29. Ibid., 34.
30. National Park Service and U.S. Department of the Interior, *Mission 66—To Provide Adequate Protection and Development of the National Park System for Human Use* (Washington, D.C.: Government Printing Office, 1956), unpaginated [14].
31. Ibid.
32. The federal influence of the NPS was forged during the Depression Era thanks to the backing of Harold Ickes, President Roosevelt's Secretary of the Interior. The NPS was given a large role as the manager of the CCC and, by 1935, was in charge of 118 CCC workgroups, which were responsible for the construction of 350 new state parks. At the height of the New Deal, the CCC accounted for 120,000 employees within the NPS.
33. Wilson, "View from the Road," 36.
34. Ibid., 34.
35. See Fraser Darling, "Man and Nature in the National Parks: Reflections on Policy," *National Parks Magazine* 43 (1969): 17, and Foresta, *America's National Parks*, 54, 61, for more on this reputation.
36. Conrad Wirth, *Parks, Politics, and the People* (Norman: University of Oklahoma Press, 1980), 278. Of course, the NPS's advocacy of modernist architecture was not limited to the efforts of Mission 66. As early as 1945, an international jury selected by the agency chose Eero Saarinen to realize the 630-foot stainless-steel arch for the Jefferson National Expansion Memorial in St. Louis.
37. *National Parks Magazine* 29, no. 120 (January–March 1955), 4, cited in Allaback, *Mission 66 Visitor Centers*, 14.
38. Conrad Wirth Papers (CWP), box 6, American Heritage Center, Laramie, Wyoming, as quoted in Allaback, *Mission 66 Visitor Centers*, 36.
39. In fact, despite the nationalistic rhetoric associated with the program of Mission 66, many of the architects hired during this spurt of high modernist visitor centers were immigrants who brought a particularly European form of modernism to the American natural landscape. See Allaback, *Mission 66 Visitor Centers*, 33, 35.
40. As a result of the National Park Service's populist view toward education and history in the 1960s, parks and historical sites became the beneficiaries of "living history" installations. Historical episodes—often embellished for dramatic effect or even fictionalized—were played out by park service actors dressed in period costume. At its most egregious, NPS rangers in red-face and various states of Native American undress traded wampum and made human pyramids. These acrobatic displays were gradually eliminated due to their excessive entertainment value, but "living history" is still considered—in the minds of the NPS Division of Interpretation—the most effective way to communicate, for example, the tension engulfing statesmen at the signing of the Declaration of Independence or the nostalgic quaintness of Colonial life in Williamsburg, Virginia. See Conrad L. Wirth, "The Mission Called 66," *National Geographic*, July 1966, 7–47.
41. Foresta, *America's National Parks*, 253.
42. Ibid., 254.
43. Other factors that hindered the realization of the visitor center included Neutra's refusal to visit the site and

his insistence upon high modernist finishes. Neutra's late color scheme ordered a change from the neutral gray that he had originally imagined to "opalescent ruby-ebony" with alternating maroon and terra cotta walls. NPS officials ultimately rejected the color scheme and painted everything in the standard "park service mustard" and in varying shades of beige.

44. The Gettysburg project has typically been dismissed as a derivative and unsophisticated example of Neutra's late work. See Willy Boesiger, *Richard Neutra, 1961–1966: Buildings and Projects* (London: Thames and Hudson, 1966), 156–65, for a description of the project, and Thomas S. Hines, *Richard Neutra and the Search for Modern Architecture: A Biography and History* (New York: Oxford University Press, 1982), 243–45, for an attempt to contextualize the project within Neutra's 1960s oeuvre.
45. Arthur Schlesinger, Jr., *The Vital Center: The Politics of Freedom* (Boston: Da Capo Press, 1949), and Serge Guilbaut, *How New York Stole the Idea of Modern Art: Abstract Expressionism, Freedom, and the Cold War* (Chicago: University of Chicago Press, 1983), 3.
46. Guilbaut, *How New York Stole the Idea of Modern Art*, 200.
47. Ibid., 4.
48. Ibid., 200. Schlesinger, Jr., *The Vital Center*, 52.
49. Schlesinger, Jr., *The Vital Center*, 6.
50. James Corner, "Eidetic Operations and New Landscapes," *Recovering Landscape: Essays in Contemporary Landscape Architecture* (New York: Princeton Architectural Press, 1999), 157.
51. U.S. Department of the Interior, *Man in Space: Study of Alternatives* (Denver: National Park Service, 1986), 8.

BEAT SPACE

1. Gilbert Millstein, "Books of the Times," *The New York Times Book Review*, 5 September 1957.
2. Jack Kerouac to Hal Chase, 19 April 1947, *Jack Kerouac: Selected Letters 1940–1956*, ed. Ann Charters (New York: Viking, 1995), 107. Five months later Kerouac wrote to Neal Cassady, "We are living at just the right time—Johnson and his London, Balzac and his Paris, Socrates and his Athens—the same thing again." Jack Kerouac to Neal Cassady, 13 September 1947, *Jack Kerouac: Selected Letters*, 130.
3. Jack Kerouac to Caroline Kerouac Blake, 25 September 1947, *Jack Kerouac: Selected Letters*, 131.
4. William S. Burroughs, "A Review of the Reviewer," *The Adding Machine* (New York: Seaver, 1986), 195.
5. Franco Moretti, *Atlas of the European Novel, 1800–1900* (London: Verso, 1998), 7. Moretti transcribes works by writers such as Jane Austen and Joseph Conrad onto maps to demonstrate how racial, social, and political conflicts are inscribed within the geography of the novel.
6. See Ann Charters's analysis of the sources for Kerouac's style in *The Beat Reader*, ed. Ann Charters (New York: Viking, 1992), 189–208.
7. During World War II, the largest one hundred American corporations received two-thirds of all wartime contracts. Consequently, the share of small business in the American economy declined by one-third. The war accelerated the process initiated by the New Deal in which the federal government grew from a miniscule apparatus to the expansive bureaucracy and military power it is today. Carroll Pursell, *The Military Industrial Complex* (New York: Harper & Row, 1972), 164–65; and Randall G. Holcombe, "The Growth of the Federal Government in the 1920s," *The Cato Journal* 16, no. 2 (Fall 1996).
8. Dennis McNally, *Desolate Angel: Jack Kerouac, the Beat Generation, and America* (New York: Random House, 1979), 51.
9. Michael Schumacher, *Dharma Lion: A Biography of Allen Ginsberg* (New York: St. Martin Press, 1992), 233.

10. Thomas McDonald, chief of the Bureau of Public Roads, asserted before Congress in 1944, "Everybody in the United States is waiting for the close of the war to get in a car and go some place," quoted in Edward Dimenburg, "The Will to Motorization: Cinema, Highways, and Modernity," *October* 73 (Summer 1995): 128. According to Dimenburg, "the highway may well be the preeminent centrifugal space of the twentieth century" (93).
11. Kerouac to Chase, 19 April 1947, *Jack Kerouac: Selected Letters*, 107.
12. Jack Kerouac, *On the Road* (New York: Viking Press, 1957), 19.
13. Leslie A. Fiedler, *The Return of the Vanishing American* (New York: Stein and Day, 1968), 16–17.
14. Kerouac, *On the Road*, 10.
15. Ibid., 17.
16. Ibid., 280.
17. Ibid., 245.
18. Henry Shryock, Jr., *Population Mobility within the United States* (Chicago: University of Chicago Press, 1964), 412.
19. Davis R. B. Ross, *Preparing for Ulysses* (New York: Columbia University Press, 1969), 160–89.
20. The idea of speed as a critical attribute of the 1950s is taken from W. T. Lhamon, *Deliberate Speed: The Origins of a Cultural Style in the American 1950s* (Washington, D.C.: Smithsonian Institute Press, 1990).
21. Kerouac, *On the Road*, 133.
22. On the development of the highway system, see Albert C. Rose, *Historic American Highways* (Washington, D.C.: American Association of State Highway Officials, 1953).
23. Jeffrey T. Schnapp, "Crash (Speed as Engine of Individuation)," *Modernism/Modernity* 6, no.1 (1999): 4.
24. Kerouac wrote to Cassady on the subject, "I wrote that book on COFFEE. . . . Benny, tea, anything I KNOW none as good as coffee for real mental power kicks." Jack Kerouac to Neal Cassady, 10 June 1951, *Jack Kerouac: Selected Letters*, 319.
25. Kathryn Shattuck, "Kerouac's 'Road' Scroll is Going to Auction," *The New York Times*, 22 March 2001.
26. Kerouac, *On the Road*, 232.
27. In his literary study of Kerouac and Burroughs, Jonathan Paul Eburne claims, "The master metaphor of [the] struggle to preserve the 'integrity' of the American subject position from the contamination by the Other was the drive to preserve the body from the corrupting influences of 'unnatural' bodily acts." Jonathan Paul Eburne, "Trafficking the Void: Burroughs, Kerouac, and the Consumption of Otherness," *Modern Fiction Studies* 43, no. 1 (1997): 62.
28. Allen Ginsberg, "Howl," in *Howl and Other Poems* (San Francisco: City Lights Pocket Bookshop, 1956), 20.
29. Kerouac, *On the Road*, 15.
30. Ibid., 274.
31. Alfred C. Kinsey, *Sexual Behavior in the Human Male* (Philadelphia: Saunders Co., 1948), 393.
32. Kerouac, *On the Road*, 119.
33. Ibid., 237.
34. Peter Hall, *Cities of Tomorrow* (Oxford: Blackwell, 1996), 294; and Kenneth T. Jackson, *Crabgrass Frontier* (New York: Oxford University Press, 1985), 238.
35. Jackson, *Crabgrass Frontier*, 4.
36. William S. Burroughs to Allen Ginsberg, 13 December 1954, William S. Burroughs, *Letters to Allen Ginsberg, 1953–1957* (New York: Full Court Press, 1982), 71.
37. Kerouac, *On the Road*, 143–44.
38. Gilbert Millstein wrote in his 1957 review, "And, finally, there is some writing on jazz that has never been equaled in American fiction, either for insight, style or technical virtuosity." Gilbert Millstein, "Books of the Times," *The New York Times Book Review*, 5 September 1957.
39. Jack Kerouac, "Essentials of Spontaneous Prose," in Charters, *The Beat Reader*, 57–58.
40. Kerouac, *On the Road*, 60.
41. Simon Sadler, *The Situationist City* (Cambridge, Mass.: MIT Press, 1999), 88–89.

42. Kerouac, *On the Road*, 106.
43. Ibid., 131.
44. An even more radical strategy for opposing the process of normalization was posed by Ginsberg. In *Howl* he incorporates a motley crew of sailors, addicts, queers, drunks, and vagabonds and their sexual deviations into the space of the city. In the sexualized public space of parks, train stations, and streets, sexual energy is exchanged:

    Who howled on their knees in the subway
    and were dragged off the
    roof waving genitals and manuscripts,
    who let themselves be fucked in the ass
    by saintly motorcyclists,
    and screamed with joy,
    who blew and were blown by those human
    seraphim, the sailors,
    caresses of Atlantic and Caribbean love,
    who balled in the morning in the
    evenings in rosegardens and the
    grass of public parks and cemeteries
    scattering their semen
    freely to whomever come who may.

    Ginsberg's provocative violation of the aesthetic norms of sexual representation was intended to release the body from its discursive confinement within the interiority of heterosexual domesticity and open it to the political space of the city.

    Ginsberg's act of sexualizing urban space was perceived as an attack on normative political power. The poet and his publisher, City Lights Books, were brought to court on obscenity charges and the book was banned. Represented by the American Civil Liberties Union, they won the case, which had the effect of bringing literature under the protection of the First Amendment. The Beat project of sexualizing urban space reveals that the conflict over free speech cannot be separated from the conflict over the body.
45. Quoted in Jackson, *Crabgrass Frontier*, 231. Homeownership in America rose from 44 percent in 1934 to 63 percent in 1972. Ibid., 215–16.
46. Kerouac, *On the Road*, 267.
47. Ibid., 28.
48. Ibid., 269.
49. David Riesman, "Autos in America," *Abundance for What? and Other Essays* (Garden City, N.Y.: Doubleday, 1964), 298–99.
50. Riesman typifies Ford's approach as ascetic, adhering to the use value of cars as engineering products, while GM's design promotes the exchange value of cars as commodities by appealing to the irrational desires of consumers. Chrysler's strategy is to cater to alternative "minority" tastes with its "feminine" design of station wagons. Kerouac employs similar attributes, assigning the Plymouth to an effeminate homosexual couple, the Cadillac to a nouveau riche gangster, while an "old thirties Ford" in its mechanical straightforwardness is employed as a nostalgic image for a lost American sincerity. Ibid.
51. Kerouac, *On the Road*, 110–11.
52. Kinsey, *Sexual Behavior in the Human Male*, 553.
53. Kerouac, *On the Road*, 203.
54. Contrary to Kerouac's misogynistic conception of gender roles, Ginsberg pursued an alternative strategy of subversive mimicry. He chose to represent the act of composing "Howl" with photographs that set it in the kitchen and bedroom of the apartment he shared with his lover, Peter Orlovsky. The kitchen utensils and unmade bed contrast with the poem's direct attack on suburban domesticity as extensions of capitalism and Cold War military build-ups:

    Moloch! Moloch! Robot apartments! invisible suburbs! skeleton treasuries! blind capitals! demonic industries! spectral nations! invincible madhouses! granite cocks! monstrous bombs!

    Thus Ginsberg's photographic mimicry of matrimonial domesticity can be read as a parody that upsets the separation of social space into private and public domains by the normative institution of heterosexual domesticity. The scandal and legal conflict over "Howl" would validate Ginsberg's position and repoliticize domesticity by taking it to court.

55. Fredric Jameson, *Postmodernism, or, The Cultural Logic of Late Capitalism* (Durham: Duke University Press, 1991), 16.
56. See Norman Podhoretz's critique of Beat literature as reactionary in comparison with the critical Bohemianism of the 1920s: "Algren hates middle-class respectability for moral and political reasons while Kerouac who is thoroughly unpolitical, seems to feel that respectability is a sign not of moral corruption but of spiritual death." Norman Podhoretz, "The Know Nothing Bohemian," *Partisan Review*, spring 1958, 307.

PORNOTOPIA

1. Hugh M. Hefner, introduction to *Inside the Playboy Mansion*, by Gretchen Edgren (Los Angeles: General Publishing Group, 1998), 6.
2. Kristin Ross, *Fast Cars, Clean Bodies: Decolonization and the Reordering of French Culture* (Cambridge, Mass.: MIT Press, 1995), 11.
3. The expression "minor architectural project" borrows from the notion of "minor literature" developed by Gilles Deleuze and Félix Guattari to describe Kafka's ability to create a "language inside another language." Deleuze and Guattari, *Kafka: Pour une littérature mineure* (Paris: Minuit, 1975), 29.
4. In fact, these oppositions regulated social and political discourse in America during the 1950s and were present as well in other magazines of the time, such as *Esquire* and *Sunshine and Health*. *Esquire*, the first male-oriented American magazine, was the most popular mainstream magazine for men during the 1940s and 1950s in the U.S., partisan of the common defense of the family and the nation. At the other extreme, *Sunshine and Health*, along with *Modern Sunbath*, were minor publications dedicated to public health and the care of the body. *Sunshine and Health* was the only American magazine that had the legal right to picture women with pubic hair during the 1950s.
5. These spaces gained recognition, in part, through the vehicle of television. The "Playboy Penthouse" TV Show aired on October 24, 1959, and ran for two seasons on Saturday nights at 11:30 p.m. on WBKB Chicago, Channel 7. This program was syndicated to a loose network of stations across the country that signed up for the show. In 1969, Hefner tried again with a similar variety show in a party setting, called "Playboy After Dark." See www.tvparty.com
6. The only exception was when Hefner caught the "Playboy Airplane" (a DC-9 equipped with a dance floor, an elliptical bed, and a sunken Roman bath), which would shuttle him between his Chicago and Hollywood homes.
7. Hefner said that only the house would enable him to "reinvent himself as a Playboy." Introduction to Edgren, *Inside the Playboy Mansion*, 11.
8. *Playboy*, December 1953, 1.
9. Hefner's friends included Arthur and Bea Paul, Eldon Sellers, and Arv Miller. In fact, the editorial address printed inside the magazine was the young Hefner's family domicile. Hefner was only twenty-two years old but had already published a self-financed comic book (Hugh Hefner, *That Toddlin' Town: A Rowdy Burlesque of Chicago Manners and Morals* [Chicago: Chi Publishers, 1951]) and had gained some experience working as a cartoonist for *Esquire* and as circulation director for the magazine *Children's Activities*.
10. The first issue of *Playboy* magazine was an immediate success. Most probably, the lure of this first issue was the nude picture of Marilyn Monroe by Tom Kelley that Hefner had bought from

John Baumgarth's Calendar Company, a society specializing in the production of pin-up pictures located in Melrose Park, on the outskirts of Chicago. By that time Monroe was already well known from the film *The Asphalt Jungle*, and the Calendar Company dared not reproduce the full nude color picture for a calendar to be sent by U.S. mail at the risk of prosecution for "obscenity." Monroe, lying on a backdrop of red velvet and peeking at the camera from behind an upraised arm, became the center of *Playboy*'s "Guide for the Urban Man." Surprisingly, Hefner was not made to appear in court. See Russell Miller, *Bunny: The Real Story of Playboy* (London: Michel Joseph, 1984), 44.

11. "Hugh Hefner." http://archive.salon.com/people/bc/1999/12/28/hefner/index1.html.
12. During the 1950s the three main dangers—"Atomic Danger," "the communist menace," and "space visitors"—seemed synonymous with exterior dangers. Basements and bunkers became the architectonic solutions provided by the American government to diminish social panic.
13. At that time, Hefner was officially still living with his wife Millie and their child, in a five-room apartment at 6052 South Harper Avenue in Chicago, although, according to his biography, he had managed to transform his office into a new residence, visiting his house less and less frequently. If the bachelor pad played with the utopian recon ciliation of work and leisure inside the domestic sphere, Hefner himself had found a way to transform his work place into his domicile. See Edgren, *Inside the Playboy Mansion*, 1–7.
14. Victor A. Lownes, cited in Miller, *Bunny*, 62.
15. See Steven Cohan, in Joel Sanders, *Stud: Architectures of Masculinity* (New York: Princeton Architectural Press, 1996), 28–41.
16. *Playboy*, September 1956, 54.
17. *Playboy*, October 1956, 65.
18. *Playboy*, September 1956.
19. *Playboy*, December 1953, 1.
20. Answering the fears of feminization and the threat of homosexuality imbedded in the return to the domestic, "The Womanization of America" by Philip Wylie, a critical reaction to the rise of the feminist movement in America, became one of *Playboy*'s constant references.
21. *Playboy*, September 1956, 54–58.
22. "*Stag Party* was to be a magazine for the bright, young, urban male who was interested in girls, fun, good living, a 'contemporary equivalent,'" Hefner would explain, "of wine, women and song although not necessarily in that order." Miller, *Bunny*, 37.
23. See Al Di Lauro and Gerald Rabkin, *Dirty Movies: An Illustrated History of the Stag Film, 1915–1970* (New York: Chelsea House, 1976); and Linda Williams, *Hardcore: Power, Pleasure, and the "Frenzy Visible"* (Berkeley: California University Press, 1989).
24. Miller, *Bunny*, 44.
25. Three semantic displacements can be identified in the shift from stag to bunny, a shift reflecting some of the characteristics of the male reader that *Playboy* would succeed in producing. First, a shift from the serious grown-up figure of the stag to the childish image of the bunny, rejuvenated, playful, and as innocent as a cuddle toy. Second, the movement from "big game" hunting to "small game" hunting. If the former aimed to hunt a unique big trophy (a wife that would last forever), the latter found pleasure in a faster hunt, the result of which would be multiple little pieces (multiple light and ephemeral sexual affairs). And third, drawing on this ambiguity between animal hunting and sexual hunting, the pair hunter-stag is replaced by the more perverse and subtle opposition between

"hunter" and "bunny," conflating not only human and animal but also sexual hunter and sexual prize. Notice that at this stage of the production of the *Playboy* logo, the bunny is the male hunter.

The well-known black and white female bunny logo—a variant of the male bunny designed in 1953—was introduced by *Playboy* in 1956 for use on derivative objects such as cufflinks, earrings, bracelets, tie tacks, and tee-shirts. After the reconstruction of the Playboy House in 1960, the logo replaced the address of the house on *Playboy*'s mailing envelopes, working as a topographic name.

26. *Playboy*, September 1956, 59.
27. Alexander von Vegesack, Peter Dumas, and Mathias Schwartz-Clauss, eds., *One Hundred Masterpieces from the Vitra Design Museum Collection*, exhib. cat. (Weil am Rheim, Germany: Vitra Design Museum, 1996), 38. I thank Alexandra Midal for her help with the research concerning design.
28. Borsani's Divan D 70 was designed between 1953 and 1954 and produced by Tecno. Represented in its "smiling" form and marked by a T, the divan became the symbol of the Italian factory. It was awarded first prize at the Tenth Triennial in Milan in 1954. *La Collection de design du Centre Georges Pompidou, Musée National d'Art Moderne—Centre de Création Industrielle* (Paris: Editions du Centre Georges Pompidou, 2001), 72–73.

    Saarinen's Tulip chair was designed in 1956 and is still produced by Knoll Associated. The characteristic feature of the Tulip series is that the supporting structure has been pared to a central stem, "like a wineglass," in order to emphasize the uniformity of table and chair and to facilitate movement.

    The chair, designed in 1947 (the first with a plastic shell), was the result of a joint attempt by Saarinen and Eames "to mold laminated wood three-dimensionally." This joint attempt resulted in the winning chair at the 1940 Organic Design in Home Furnishings competition, held by the Museum of Modern Art in New York. Continued design development by Saarinen finally led to the Womb Chair of 1947. Ibid.
29. *Playboy*, September 1956, 57.
30. *Playboy*, October 1956, 67–68.
31. *Playboy*, September 1959, 59–60.
32. In continuity with the tradition of the "stag," hunting would become one of the constant subjects of *Playboy*. In March 1958 the magazine published "The Right Honourable Hide," which illustrated a selection of hunting accessories for the urban Playboy. A slight transformation has taken place: the hunting weapons have become ornamental objects, souvenirs coming from a colonial safari, that now cover the wall of the bachelor apartment; a design chair takes the place of the horse; a mini-bar replaces water and food provisions; and the balls of the portable casino game replace munitions. *Playboy* introduces the young urban male to indoor hunting: "It can tote your whisky, keep your ice cubes frosty, offer you a spot to sit down, protect your Francotte shotgun, cart your pipe cool, your cigarette firm, your feet dry, your money crisp and your pants in place." *Playboy*, March 1958, 56.
33. *Playboy*, September 1956, 59.
34. Note, for instance, the resemblance of the atmosphere of the Playboy Penthouse Apartment and the domestic killing machine described in the 1990s in Bret Easton Ellis' *American Psycho* (New York: Vintage Books, 1991), 217, 290, 304–5, 344.
35. See Dolores Hayden, *Gender, Housing and the Family Life* (Cambridge, Mass: MIT Press, 1981), and Dolores Hayden, *The Grand Domestic Revolution: A History of Feminist Designs for American Homes, Neighborhoods and Cities* (New York: Norton, 1984).
36. *Playboy*, September 1956, 60.

37. See Adrian Forty, *Objects of Desire* (New York: Pantheon, 1986).
38. *Playboy* does not hesitate to advise the bachelor to hire a maid once a week to finish the cleaning activities, fearing the possibility of a woman taking back these duties. *Playboy*, September 1956, 60.
39. *Playboy*, October 1956, 70.
40. *Playboy*, September 1956, 60.
41. Ibid.
42. *Playboy*, October 1956, 70.
43. Hugh Hefner, "Editorial," *Playboy*, November 1956, 2.
44. *Playboy*, January 1959, 7.
45. Hugh Hefner, cited in Gretchen Edgren, *Playboy, 40 ans*, trans. Jacques Collin (Paris: Editions Hors Collection, 1996), 7.
46. Miller, *Bunny*, 56.
47. Cited in Hugh M. Hefner, ed., *The Twelfth Anniversary Playboy Cartoon Album* (Chicago: Playboy Press, 1965), 22.
48. Miller, *Bunny*, 55.
49. "The Varga girl" was first commercialized in *Esquire* (which eliminated the final "s" from the name of its creator) through the production of calendars in 1940. Charles G. Martignette and Louis K. Meisel, *The Golden Age of the American Pin-Up* (New York: Taschen, 1999), 26–27.
50. Beatriz Colomina has made manifest the relationship between X-ray techniques and regimes of visibility within the modern house. Beatriz Colomina, "The Medical Body in Modern Architecture," in *AnyBody*, ed. Cynthia Davidson (Cambridge, Mass.: MIT Press, 1997), 228–38.
51. Michel Foucault, *Histoire de la sexualité, 1. La volonté de savoir* (Paris: Gallimard, 1976), 43–45.
52. See the pin-up drawing by Edward D'Ancona in Charles G. Martignette and Louis K. Meisel, *L'Age d'or de la pin up Américaine* (Paris: Taschen, 1996), 112.
53. Miller, *Bunny*, 57.
54. Ibid.
55. John Berger, *Ways of Seeing* (New York: Penguin Books, 1977), 54; Laura Mulvey, *Visual and Other Pleasures* (Bloomington, Ind.: Indiana University Press, 1989). See especially her use of the notion "male gaze" in "Visual Pleasure and Narrative Cinema," *Screen* 16, no. 3 (Autumn 1975): 6–18.
56. In 1948 only 64 percent of all homes had telephones; by 1963, 83 percent had phones.
57. For instance, Miss April 1955 was pictured lounging on a charcoal-grey sofa in a pair of checkered matador pants and nothing else, with Hefner's pipe featured prominently in an ashtray nearby. Miss November 1955 posed nude with a towel wrapped loosely around her, while Hefner's tie was shown draped over the bathroom mirror.
58. Hundreds of letters arrived at *Playboy's* headquarters praising the project and asking how and where to find the objects that furnished the apartment. See Edgren, *Playboy, 40 ans*, 36.
59. Miller, *Bunny*, 34.
60. *Playboy*, May 1959, 50–53. By 1959, maybe as a sign of America's leaving behind the times of the McCarthy "commies and queers hunt," or maybe as "the antidote effect" of the same policy, *Playboy*'s circulation hit one million, overtaking *Esquire*. The 1950s and early 1960s were not only the years of the rigors of the Cold War but also the golden age of *Playboy*.
61. Edgren, *Inside the Playboy Mansion*, 8.
62. Ibid.
63. Miller, *Bunny*, 76.
64. Hefner's relationship to the Chicago Mafia is unclear. Accounts differ within the publisher's various biographies and the histories of the *Playboy* Empire.
65. A more careful look at the plan and section indicates that the rotating bed might have been added to the drawings later, perhaps after Hefner's own bed was constructed. *Playboy*, June 1962. See also *Chicago Daily News*, 23 September 1959.
66. The ad reads, "Porsche—the car that 'lets you lead a double life.'" *Playboy*, June 1962, 49.

67. A study of the similarities between the Playboy House and the house built by Charles and Ray Eames, their "Playhouse" (both being structured around notions of play, entertainment, the combination of living and working spaces, and the fact that it, like a multimedia production), could be extremely fruitful. The framework of this paper does not allow me to develop such an investigation here, but it will be pursued in a future paper.
68. *Playboy*, January 1960, 47.
69. Edgren, *Playboy, 40 ans*, 33.
70. Miller, *Bunny*, 52, and *Playboy*, January 1959, 69.
71. Miller, *Bunny*, 52. See also *Playboy*, January 1960, 47–48.
72. Miller, *Bunny*, 85.
73. *Playboy*, January 1960, 47.
74. Ibid., 48.
75. The first television stage for "Playboy's Penthouse" was, in fact, an imitation of the interior of the Playboy House's hall.
76. Miller, *Bunny*, 10. The issue of the "black guests" was the cause of an intense controversy among the show's staff, although most of the African-American guests were well-known musicians (such as Ray Charles and Sammy Davis Jr.) invited to amuse the white crowd. Otherwise, African Americans were not guests but rather waiters. Not a single Playmate was African-American until 1965. See Edgren, *Playboy, 40 ans*, 88–89.
77. Michel Foucault, "Des espaces autres," *Dits et écrits, 1954–1988* (Paris: Éditions Gallimard, 1994), II: 752–62.
78. Once again, I use the language of Foucault. He distinguished between the "crisis heterotopias"—such as the sacred or forbidden spaces of the so-called "primitive" societies, inhabited during a "state of crisis" by adolescents, pregnant women, etcetera—and the modern "heterotopias of deviation"—"those in which individuals whose behavior is deviant in relation to the required mean or norm are placed." Ibid., 758–60.
79. Ibid., 759.
80. The treatment of "orientalized" and "primitive" motifs in *Playboy* deserves additional study. Although the word "harem" is never explicitly used in the magazine, the representation of nude women as a tribe or harem that belonged to the "colonial" Hefner is persistent. To be able to understand the close relationship between pornography and colonial representation, it is important to remember that the first photographs of exposed breasts printed in a color were published in *National Geographic*, which was able to publish pictures of the nude women (and very rarely men) of "primitive tribes" without being accused of sexual exploitation or obscenity. Here the distinction between dressed and undressed served to articulate the difference between the civilized and the primitive, the human and the animal, making privacy and clothing a privilege of occidental and developed societies.
81. This was similarly the case in the design of the water-divided penthouse by Donald Jaye.
82. It seems evident that the Playboy House was used as a film studio, both during the 1960s and later, as inspiration for the sites of different remakes, such as *Playboy Pajama Parties* (1982, 1996) and *Girl-Next-Door* (1975, 1983, 1997), but I have not found explicit information related to this.
83. Louis Marin, "Sur la création de l'Île d'Utopie," *Utopies: Jeux d'espaces* (Paris: Les Éditions de Minuit, 1973), 140.
84. For a discussion of the competition between furniture preferences during the postwar period, see Mary Davis Gillies, ed., *What Women Want in Their Bedrooms of Tomorrow: A Report of the Bedroom of Tomorrow* (New York: McCall Corporation, 1944). For a critical analysis about the dispute between the double bed and the twin beds, see Jeannie Kim, "Sleep With Me" (unpublished paper, Princeton University, Fall 2000).

85. See Edgren, *Playboy, 40 ans*, 2–3. Comparing different sources, it is unclear whether the round rotating bed was the one designed by Donald Joyce in 1959 and published in the May 1962 issue of *Playboy*. The different hagiographical histories of *Playboy* designate Hefner as the "creator" of the bed (once more depicting the publisher as designer and architect).
86. Hefner's bed surpassed in notoriety Cary Grant's and Tyrone Power's square beds, and Lana Turner's heart-shaped bed.
87. Miller, *Bunny*, 8–9. The writer Tom Wolfe provides a better explanation for Hefner's attachment to the rotating bed: "The bed touring around its own orbit, and, on the right, the Ampex video cassette playing on the screen something beautiful that has just happened inside the same bedroom, Hefner is at the center of the world, repeated at every turn of the bed, at the center of a controlled world, where he is the only king, without ever being ejected, but always plunged into.... It is the perfect rotation.... After every turn, the nirvana, the ambrosia, here, at the center so everybody can see it, Playboy's headlight." Cited in Miller, *Bunny*, 9.
88. Miller, *Bunny*, 1.
89. Mimicking Hefner's childish etiquette, the legendary Friday night parties at the Playboy House became "Pajama Parties" in which guests wore only smoking jackets, night gowns, and underwear. Edgren, *Inside the Playboy Mansion*, 66–67.
90. Miller, *Bunny*, 20.
91. Siegfried Gideon, *Mechanization Takes Command* (New York: Oxford University Press, 1948); Henri Lefebvre, *Position: Contre les technocrates* (Paris: Gauthier, 1967).
92. *Playboy*, January 1958, 29, 36.
93. Edgren, *Playboy, 40 ans*, 32.
94. David Fernbach, *A Theory of Capitalist Regulation: The U.S. Experience* (London: New Left Books, 1976), 110.
95. Before becoming a playmate, Dugan was known for her role in the 1956 film *The Great Man* by Jose Ferrer (Universal), in which she played "the network receptionist who 'made a great exit' from Keenan Wynn's apartment." In 1957 she appeared in the TV series "And-Away-We-Go-Girls" with Jackie Gleason.
96. *Playboy*, July 1957, 60.
97. Ibid., 61.
98. *Playboy*, January 1960, 47.
99. Foucault, "Des espaces autres," II: 760. Foucault names both boarding schools and brothels in his list of the heterotopias within modern society.
100. Miller, *Bunny*, 9.
101. In each dormitory, the most experienced playmate, designated as the "Bunny Mother" and presented to the other girls as counselor and adviser, was actually in charge of maintaining the bunnies' "quality" in her dorm. She rewarded Bunnies' "merits" for achievements (such as serving a large number of drinks and good maintenance of their bunny costumes) with "weekly awards" and punished their faults (such as chewing gum, having messy hair, bad nails, bad make-up, using "profane language," etc.) with the threat of their being expelled from the house. Ibid., 10–11.
102. Ben Hecht, "No Room for Vice," *Playboy*, January 1959, 51–54.
103. Ibid., 54.
104. Chicago was one of the main centers of prostitution in America during the early twentieth century. Between the late 1920s and the 1950s, the Mafia (headed first by Jim Colisimo and later by Al Capone) controlled all spaces of prostitution. As Captain Golden explained in the *Playboy* article itself: "I can remember innumerable such centers.... They were very expensively furnished and the ladies wore ball gowns.... Beginning in 1890, Chicago experienced a brothel boom unique in the Republic. Hundreds of madams from all over came sashaying into the great cow-killing metropolis, bringing their full staffs with them, including pimps and piano professors. Thousands of venturesome young farmers' daughters

and prairie village maidens came pouring out of the Santa Fe and the La Salle Street stations on their own. A few weeks later found them scented and silken-gowned in the Tenderloin havens. Around 1900 Chicago was the Republic's unchallenged center of bawdry. New Orleans, New York and San Francisco were unhappy runners-up. Vice in those communities was a glittering side-line activity. In Chicago, it was half of the town. Crystal chandeliers blazed in every third parlor on the Near North and the Near South Sides. Pimps canvassed the city's office buildings like crack salesmen. Most of the city's cafés, theatres and rallying places were barred to respectable females. Only bawds enjoyed the town. On a Saturday night the tune-filled bordellos were as jam-packed as are the city's beaches today." *Playboy*, January 1954, 51. See also Vern L. and Bonnie Bullough, *Women and Prostitution* (Buffalo, N.Y.: Prometheus Books, 1987).

105. It has yet to be investigated how architecture participates in the construction of sexuality, and conversely, how different sexual regimes require the production of different architectonic devices. This is nothing more than a variation of the question that Beatriz Colomina and Mark Wigley have already posed differently in various articles and books about the relationship between gender, sex, and architecture: "the relationships between the role of gender in the discourse of space and the role of space in the discourse of gender." Mark Wigley, "Untitled: The Housing of Gender," in *Sexuality and Space*, ed. Beatriz Colomina (New York: Princeton Architectural Press, 1992), 327–89.
106. This feedback relationship between the house and the magazine set up a model for future pornographic industries such as Larry Flynt's *Hustler*.
107. Miller, *Bunny*, 81.

ILLUSTRATION CREDITS

2 *House Beautiful*; **3** *Life*; **10** FoundImage; **13** The Academy of Motion Picture Arts and Sciences Archives; **15** *House Beautiful*; **16 left** Photography Archive, Harvard University Art Museum, photographer Joe Steinmetz; **16 right** National Archives; **20** *Life*; **23 top** U.S. Air Force; **23 bottom** Courtesy of Herman Miller, Inc.; **27** U.S. Air Force; **29–30** Clarence S. Yakum and Robert M. Yertes, *Army Mental Tests* (New York: Henry Holt and Company, 1920); **41–43** U.S. Air Force; **45** Courtesy of Herman Miller, Inc.; **47** Courtesy of Herman Miller, Inc.; **51–52 top and bottom** General Dynamics Corporation, U.S. Air Force; **55 left** Aluminum Company of America, *Process Industries Applications of Alcoa Aluminum*, company catalog, 1955; **55 right** *Life*; **57** *Saturday Evening Post*; **59** *Newsweek*; **64** *Life*; **71** Aluminum Company of America, *Process Industries Applications of Alcoa Aluminum*, company catalog, 1955; **75** *The New Yorker*; **77, 78–80** *The New Yorker*; **86–90** John Harper, *Design Forecast* (Aluminum Company of America brochure, Dec. 1965), Library and Archives Division, Historical Society of Western Pennsylvania, Pittsburgh, Penn., Alcoa Archives, Marketing and Sales Series, MSS Box no. 138, folder #5; **91** Courtesy of Monsanto Company, **94** Courtesy of Monsanto Company; **97 left and right** *Life*, January 1946; **99** Courtesy of Monsanto Company; **111** Courtesy of MIT Department of Architecture; **113 top and bottom** Courtesy of *Modern Plastics*; **115** Courtesy of Monsanto Company; **117** Courtesy of Monsanto Company; **119** Courtesy of Monsanto Company; **120 top and bottom left and right** Courtesy of Monsanto Company; **121–23** Courtesy of Monsanto Company; **127** Gil Asakawa and Leland Rucker, *The Toy Book* (New York: Alfred A. Knopf, 1991); **129** "Planning for Complete Flexibility," *Architectural Forum*, April 1950; **130** "Lessons from Long Island," *Builder*, December 1988; **132** "Planning for Complete Flexibility," *Architectural Forum*, April 1950; **134 top** "Builder Operations," *Architectural Forum*, April 1950; **134 bottom** "Tailor-Made Houses," *Architectural Forum*, April 1950; **136–37** "Building Type Studies No. 161 . . . Houses," *Architectural Record*, May 1950; **139 top and bottom, 140** "Six Moon Hill," *Architectural Record*, June 1950; **143** *Life*, July 16, 1951; **145 top** Eames Office, TY.MC 5025/d; **145 bottom** Library of Congress, Prints and Photographs Division, Lot 13176-1, No. 9; **147 left** National Air and Space Museum, Smithsonian Institution; **147 right** Eames Office, EH.HM 5045/d; **148** *Architectural Forum*,

September 1950; **149** *Manual of Instruction, The Mysto Erector* (New Haven, Conn.: A. C. Gilbert Co., 1915); **151** Robert Snyder, *R. Buckminster Fuller: An Autobiographical Monologue/Scenario* (New York: St. Martin's Press, 1980); **155** R. Buckminster Fuller, *Tetrascroll, Goldilocks, and the Three Bears* (New York: St. Martin's Press, 1982); **156 top** Konrad-Wachsmann Archive, AdK, Berlin, USAF—Hangar; **156 bottom, 158** Konrad Wachsmann, *The Turning Point of Building* (New York: Reinhold Publishing Corporation, 1961); **159–60** *Popular Science Monthly* LXIV, Nov. 1903; **161** National Air and Space Museum, Smithsonian Institution; **162** *National Geographic Magazine* 110, Aug. 1956; **163** Konrad Wachsmann, *The Turning Point of Building* (New York: Reinhold Publishing Corporation, 1961); **165** Eames Office, TY.MQ 5024/d; **167** *Life*, July 16, 1951; **168, 170 left,** Department of Forest Economics; **170 right** National Park Service Historic Photograph Collection, Harpers Ferry Center; **173** Liberty Bell, National Park Service, and family, photographer Walter H. Miller; **175** Department of Forest Economics; **176** National Park Service Historic Photograph Collection, Harpers Ferry Center, photographer Jack E. Boucher; **179** U.S. Department of the Interior; **181–82, 185 left and right** Collection of the Athenaeum of Philadelphia, photographer Lawrence S. Williams, Inc.; **188** U.S. Department of the Interior, *Man in Space: Study of Alternatives* (Denver: National Park Service, 1986); **190–91** Map, U.S. Department of the Interior, and overlay, Roy Kozlovsky; **201** Lisa Phillips, *Beat Culture and the New America, 1950–1965* (New York and Paris: Whitney Museum of American Art in association with Flammarion, c. 1995); **203** Getty Images; **205** Alfred C. Kinsey, *Sexual Behavior in the Human Male* (Philadelphia: Saunders Company, 1948); **213** Carolyn Cassady.